J. Morrow, Math Dept.

Differential and Riemannian Geometry

DIFFERENTIAL AND RIEMANNIAN GEOMETRY

by
DETLEF LAUGWITZ
Professor of Mathematics
Technische Hochschule Darmstadt
Germany

Translated by
Fritz Steinhardt
The City College
The City University of New York
New York

1965

ACADEMIC PRESS
NEW YORK AND LONDON

ACADEMIC PRESS INC.
111 Fifth Avenue, New York, New York 10003

United Kingdom Edition published by
ACADEMIC PRESS INC. (LONDON) LTD.
Berkeley Square House, London W.1

Library of Congress Catalog Card Number: 64-21670

PRINTED IN THE UNITED STATES OF AMERICA

First published in the German language under the title *Differentialgeometrie* and copyrighted in 1960 by B. G. Teubner, Verlagsgesellschaft, Stuttgart.

Preface to the German Edition

This textbook is intended to be an introduction to classical differential geometry as well as to the tensor calculus and to Riemannian geometry. The material it contains should fit into a two-semester course at the upper college or first-year graduate level. A systematic development has been aimed at, but the author also believes hopefully that none of the more important special theorems will be found missing. To be sure, Chapter I on the Theory of Curves serves mainly as a preparation for the later chapters. Thus the reader interested especially in the theory of curves for its own sake is also referred to any of the numerous textbooks on this particular subdiscipline.*

The tensor calculus has been used consistently. Its notation is introduced even in the theory of surfaces (Chapter II), so that for the subsequent definition of tensors and their operations (in Chapter III), concrete instances are immediately available. The advantages of the tensor calculus lie in its great generality, in the greater transparency of its notation as compared to the Gaussian notation still in frequent use, and in the immediate applicability to the special problem of formulas written in tensor notation. The author does not agree with the frequently voiced objection that equations in tensor notation are not "written in invariant form" (see III 8.5). For lack of space, no account is given of the calculus of differential forms, which is an occasionally useful but not sufficiently general abbreviated form of a part of the tensor calculus. However, anyone who has acquired a sufficient familiarity with the tensor calculus will be able to handle this.

Also in Chapter III, we treat the elements of Riemannian geometry, which owes its existence to the requirements of a systematic development of the intrinsic (inner) geometry of surfaces. Consequently there is still much material on the theory of surfaces to be found in Chapter III, as well as in the following Chapter IV. In the latter will be found some results that have not previously been available in textbooks, or in book form generally.

---

* For instance, Volume I of the two-volume work by G. Scheffers: "Anwendungen der Differential- und Integralrechnung auf Geometrie," 3rd ed., Berlin and Leipzig, 1923; or the works of J. Edwards: "Differential Calculus," 2nd ed., London, 1896, and Volume I of his treatise on the "Integral Calculus with Applications," 2nd ed., London, 1930 (reprinted, New York, 1954).

With these it is hoped that the reader will be brought in contact with some areas of present-day research (to be sure, the selection had to be a subjective one) while at the same time gaining a deepened understanding of the theory of surfaces and of Riemannian geometry. In presenting topics from the global theory (Chapter V), the author also has kept the systematic point of view to the fore, and has chosen results from among the wealth of material mainly with an eye to methodical coherence.

References to the literature have deliberately been kept on the sparse side. Apart from some especially important papers, original sources are in general cited only if they are not treated in one of the following three encyclopedic standard works:

"Enzyklopädie der mathematischen Wissenschaften," Volume III, 3, Leipzig, 1902 to 1927 (particularly for Chapters I and II);

J. A. Schouten: "Ricci-Calculus," 2nd ed., Berlin, 1954, Springer Grundlehren Series, Volume 10 (for Chapters III and IV);

N. W. Efimov: "Flächenverbiegung im Grossen," with an appendix by E. Rembs and K. P. Grotemeyer, Berlin, 1957 (for Chapter V).

We also refer to the short historical sketch of the subject in the Appendix of the present book. The Appendix further contains three prerequisite results from Analysis, and a list of some of the most important formulas.

Passages in small print may be left out on a first reading. The book contains 154 exercises which the reader should be able to tackle after studying the sections where they are placed. Those whose solution depends on more than an application of the main text have been provided with hints that are usually quite detailed. A few exercises requiring new ideas can be spotted by noting that they contain references to original papers.

An effort has been made to avoid the well-known incorrect type of argument* found in classical differential geometry. On the other hand, one must avoid smothering the essential geometry with the analytical apparatus. We hope to have steered a course in between Scylla and Charybdis!

The author owes much to the rich teaching experience of Professors G. Lyra (Göttingen) and F. Löbell (Munich). More than a few points of novelty in subject matter or exposition are derived from their lecture courses or from discussions in the circle of geometers who meet every

* Such as the following typical "Theorem," often used implicitly: "A function either vanishes everywhere, or it vanishes nowhere; however if it does have a zero at a point, then its derivative at this point is certainly not zero." (The author is indebted to H. Kneser for this formulation.)

October at Oberwolfach. A note of gratitude is due here to the leader of those sessions for many years, Professor W. Süss (deceased 1958).

The author wishes to express his thanks to Professor G. Köthe, the editor of this series of textbooks, who asked him to write this book. He also wants to record his particular appreciation to the editor, and to the publishers as well, for many instances of unstinting cooperation. Cordial thanks are due also to the author's colleagues W. Barthel (Saarbrücken) and K. Jörgens (Heidelberg) for important comments in proofreading, and to Mr. H. Böttcher (Darmstadt) for his help in preparing the drawings.

Darmstadt
Summer 1960

DETLEF LAUGWITZ

Translator's Preface

In this English translation of Professor Laugwitz' *Differentialgeometrie*, a number of minor errors and misprints in the German original have been corrected.

Even the expanded English title of this book does not give adequate indication of its scope. It leads from classical differential geometry through tensors, differentiable manifolds, Riemannian geometry, and linear connections to a broad vista from the viewpoint of general metric differential geometry; the presentation is always by means of concrete geometric problems. While there is no lack of recent high-level treatments in English of Differentiable Manifolds or Foundations of Differential Geometry, one cannot readily find an equally well-organized ascent to the heights.

November, 1964 FRITZ STEINHARDT

Author's Note

The text of this English edition is changed very little from that of the original German edition. A number of minor slips and printing errors have been corrected. I am grateful to many readers for pointing these out, especially to my colleagues Heinz Hopf, F. Flohr, K. Voss, and last but not least, F. Steinhardt. A few references to recent publications have been added.

I would also like to express my thanks to Dr. F. Steinhardt for his excellent translation, and to him as well as to Academic Press for their helpful cooperation.

DETLEF LAUGWITZ

Contents

CHAPTER V

SELECTIONS FROM DIFFERENTIAL GEOMETRY IN THE LARGE

APPENDIX I

APPENDIX II

SOME PREREQUISITE THEOREMS OF ANALYSIS

APPENDIX III

SUMMARY OF FORMULAS

CHAPTER I

Local Differential Geometry of Space Curves

§ 1. Differential-Geometric Properties of Curves

1.1. Parametric Representation

A space curve may be thought of as having been generated by the motion of a point. With this idea in mind one may regard the letter t, in what follows, as a time parameter.

Definition. *A curve in three-dimensional Euclidean space is given by* $\mathfrak{x} = \mathfrak{x}(t)$. *Here* $\mathfrak{x} = (x_1, x_2, x_3)$ *is the position vector and the parameter* t

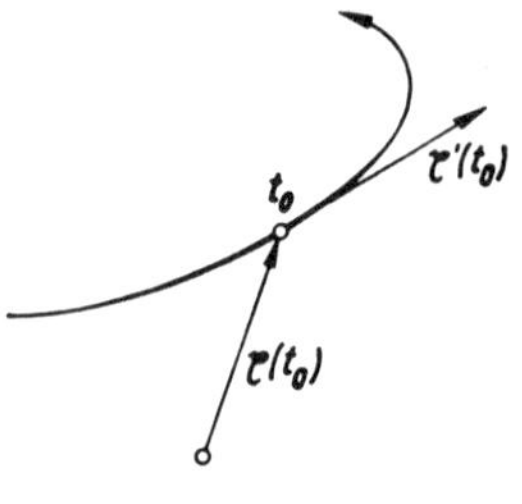

FIG. 1. Curve and tangent line.

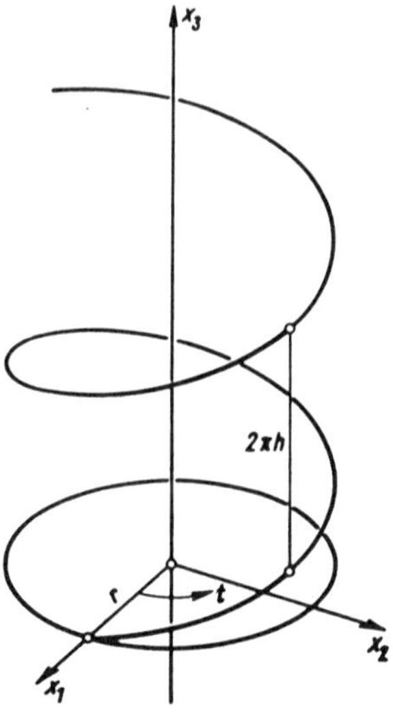

FIG. 2. Circular helix.

ranges over an open interval, $a < t < b$. *The cases* $a = -\infty$ *or* $b = +\infty$ *are admissible possibilities.* The three functions $x_i(t)$ $(i = 1, 2, 3)$ will be subject to requirements of differentiability sufficient for the purposes of differential geometry. In what follows we shall always assume that these functions $x_i(t)$ have continuous derivatives up to and including the third order. As a further assumption concerning curves and their parametrizations in what follows, we shall require that, for all t, the vector

$$\dot{\mathfrak{x}}(t) = \frac{d\mathfrak{x}}{dt} \tag{1.1}$$

not be the null vector. Geometrically this means, for $t = t_0$, that the equation

$$\mathfrak{x} = \mathfrak{x}(t_0) + \lambda \cdot \dot{\mathfrak{x}}(t_0), \qquad -\infty < \lambda < \infty \tag{1.2}$$

represents the tangent line to the curve $\mathfrak{x} = \mathfrak{x}(t)$ at the point $\mathfrak{x}(t_0)$ (see Fig. 1).

We give examples of parametric representations of some important curves:

1. For the *straight line* through the end point of a vector $\mathfrak{a}$ and in the direction of a nonzero vector $\mathfrak{b}$:

$$\mathfrak{x}(t) = \mathfrak{a} + t \cdot \mathfrak{b}, \qquad -\infty < t < \infty.$$

2. For the *circle* with center at (the end point of) $\mathfrak{m}$, with radius $r > 0$, and lying in the plane spanned by the mutually perpendicular unit vectors $\mathfrak{e}_1$, $\mathfrak{e}_2$:

$$\mathfrak{x}(t) = \mathfrak{m} + r\mathfrak{e}_1 \cos t + r\mathfrak{e}_2 \sin t, \qquad 0 \leqq t < 2\pi.$$

3. For the *circular helix* with the x_3-axis as its axis (see Fig. 2):

$$x_1(t) = r \cdot \cos t; \qquad x_2(t) = r \cdot \sin t; \qquad x_3(t) = h \cdot t,$$

where r and h are positive constants. The number $2\pi h$ is called the *pitch* of the helix.

1.2. Arc Length

We know from the calculus that the arc length of a curve from $\mathfrak{x}(t_0)$ to $\mathfrak{x}(t)$ is given by the formula

$$s(t) = \int_{t_0}^{t} \sqrt{\dot{\mathfrak{x}}^2}\, dt. \tag{1.3}$$

It seems natural to introduce the arc length s as a new parameter for our space curve. Before we proceed to do this, however, we had better investigate under what conditions it is permissible to change to a new parameter. We must require of such a *parameter transformation* that the new parametrization should be, like the old, in accord with the properties stipulated in the definition of space curve. This will be the case if the new parameter $\bar{t}$ is a function of class[1] C^3 of the old,

$$\bar{t} = \bar{t}(t), \qquad \text{with}^2 \qquad d\bar{t}/dt \neq 0, \tag{1.4}$$

and is such that the inverse function $t = t(\bar{t})$ is likewise of class C^3, for under these conditions, $\bar{t}(t)$ is a monotonic function. Since

$$\bar{\mathfrak{x}}(\bar{t}) = \bar{\mathfrak{x}}(\bar{t}(t)) = \mathfrak{x}(t) = \mathfrak{x}(t(\bar{t})), \tag{1.5}$$

the curve is now parametrized in terms of the parameter $\bar{t}$; the differentiability requirements of our definition of curve are satisfied, as is shown by (1.5) in conjunction with the chain rule; and, in particular, it follows from (1.5) that

$$\frac{d\bar{\mathfrak{x}}}{d\bar{t}} = \frac{d\mathfrak{x}}{dt} \cdot \frac{dt}{d\bar{t}},$$

so that the first derivative of $\bar{\mathfrak{x}}$ with respect to $\bar{t}$ is not the null vector. We are now in a position to verify that the transition to arc length s is an allowable parameter transformation. As to the differentiability conditions, we leave their verification to the reader (see Exercise 1.2.1). Further, we have

$$\frac{ds}{dt} = \sqrt{\dot{\mathfrak{x}}^2} = |\dot{\mathfrak{x}}|,$$

so that ds/dt is not zero. If the parameter t is the arc length s, we denote parametric derivatives with respect to s by primes $'$ and we see that

$$\mathfrak{x}'^2 = 1. \tag{1.6}$$

Thus $\mathfrak{x}'$ is a unit vector. If we interpret the equation $\mathfrak{x} = \mathfrak{x}(t)$ kinematically (as we did at the beginning of this section) by interpreting t as time, then $\dot{\mathfrak{x}}(t)$ is the velocity vector. If $t = s$ then the magnitude of this velocity vector is 1.

[1] We recall that a function is said to be of class C^r if it has continuous derivatives up to and including those of order r.

[2] In geometry, just as in mechanics and physics, it is often convenient to use the same letter for the functional symbol and for the dependent variable, as in $\bar{t} = \bar{t}(t)$. We shall not hesitate to follow this usage.

Exercise 1.2.1. Show that $s = s(t)$ is an allowable parameter transformation.

Exercise 1.2.2. Find parametrizations in terms of arc length s as parameter, for: (a) the straight line, (b) the circle, (c) the circular helix.

Exercise 1.2.3. Prove that if $|\dot{\mathfrak{x}}| = 1$ holds for a parameter t, then $t = \pm s + \text{const}$.

1.3. Geometric Properties

On account of its geometric significance, the parameter s plays a special role and is sometimes called the natural parameter. Let us investigate in a more general way which of the properties of a curve $\mathfrak{x}(t)$ might be called true geometric properties. They must be properties that depend only on the *points* of the curve and that do not depend on the manner in which these points are represented. Such properties must not change if a different representation of the points of the curve is chosen. Thus they must be *invariant* under

(a) transformations of the coordinate system in space,

(b) allowable parameter transformations of the curve.

By writing our formulas in vector notation, we will be sure that the properties they express are invariant in the sense of (a); this is because entities expressed in vector notation are evidently independent of any particular choice of the orthonormal coordinate trihedron. To obtain invariance in the sense of (b) also, we shall use the natural parameter, namely, arc length, in the parametrization of curves.

1.4. The Moving Trihedron

In line with what has just been said, we shall assume from now on that the curve is given in terms of arc length s as the parameter, thus: $\mathfrak{x} = \mathfrak{x}(s)$. Then by Eq. (1.6), we have

$$\mathfrak{x}'^2 = 1,$$

from which it follows, by differentiation with respect to s, that

$$\mathfrak{x}'\mathfrak{x}'' = 0. \tag{1.7}$$

Let us assume for the moment that $\mathfrak{x}'' \neq 0$ and let us introduce the unit vectors

$$\mathfrak{v}_1(s) = \mathfrak{x}', \qquad \mathfrak{v}_2(s) = \frac{\mathfrak{x}''}{|\mathfrak{x}''|} \tag{1.8}$$

We complement this pair of vectors by introducing

$$\mathfrak{v}_3(s) = \mathfrak{v}_1 \times \mathfrak{v}_2 \qquad (\text{hence, } |\mathfrak{v}_3| = 1), \tag{1.9}$$

thus obtaining an orthonormal trihedron, for which

$$\mathfrak{v}_i(s)\mathfrak{v}_k(s) = \delta_{ik}. \tag{1.10}$$

Here δ_{ik} is the Kronecker delta, defined by

$$\delta_{ik} = \begin{cases} 1 & \text{if} \quad i = k \\ 0 & \text{if} \quad i \neq k. \end{cases}$$

The vectors $\mathfrak{v}_2(s)$, $\mathfrak{v}_3(s)$ are perpendicular to the tangent direction at the point $\mathfrak{x}(s)$, since $\mathfrak{v}_1(s)$ is a unit vector in this tangent direction. The vector $\mathfrak{v}_1$ is called the *unit tangent vector*; the vector $\mathfrak{v}_2$ is called the *unit principal normal*; the vector $\mathfrak{v}_3$ is called the *unit binormal*. Considered as local vectors with initial point $\mathfrak{x}(s)$, the three vectors $\mathfrak{v}_i(s)$ are called the *moving trihedron* of the curve (see Fig. 3).

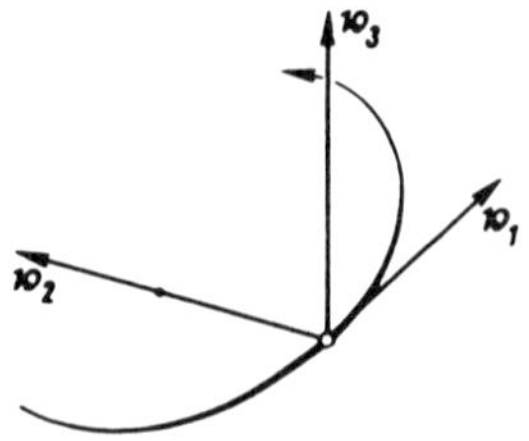

FIG. 3. The moving trihedron.

The moving trihedron is defined only at points of the curve where $\mathfrak{x}''(s) \neq 0$. If this vector vanishes only at isolated points of the curve, then the investigations that follow are restricted to the pieces of the curve between such isolated points. The only curves that will be entirely excluded by the restriction $\mathfrak{x}''(s) \neq 0$ will therefore be the curves for which $\mathfrak{x}''(s) = 0$ holds for all s. But in such a case, integration yields $\mathfrak{x} = s\mathfrak{a} + \mathfrak{b}$, where $\mathfrak{a}$ and $\mathfrak{b}$ are constant vectors, and these curves are precisely all the straight lines.

The vector $\mathfrak{x}''(s) = \mathfrak{v}_1'(s)$ measures the rate at which the unit tangent vector changes as the curve is traversed at uniform speed ($= 1$, since $t = s$ and $|d\mathfrak{x}/ds| = 1$). If this rate of change is of large magnitude, then the curve changes direction rapidly. The number

$$\kappa(s) \underset{\text{def}}{=} |\mathfrak{x}''(s)| \tag{1.11}$$

is called the *curvature* of the curve at the point $\mathfrak{x}(s)$. It is a measure of how much the curve deviates from a straight-line course. The condition $\kappa(s) \equiv 0$ characterizes the straight lines.

1.5. The Formulas of Frenet

For the other two vectors $\mathfrak{v}_2$, $\mathfrak{v}_3$ of the moving trihedron, their rates of change $\mathfrak{v}_2'$, $\mathfrak{v}_3'$ are obtained as follows: Since the three vectors $\mathfrak{v}_i(s)$ are linearly independent, we can write

$$\mathfrak{v}_k' = \sum_{i=1}^{3} a_{ki}\, \mathfrak{v}_i \qquad (k = 1, 2, 3), \tag{1.12}$$

where the coefficients $a_{ki}(s)$ are still to be determined. Forming the dot product of (1.12) by $\mathfrak{v}_j$, we obtain

$$\mathfrak{v}_k'\, \mathfrak{v}_j = \sum_{i=1}^{3} a_{ki}\, \delta_{ij} = a_{kj}. \tag{1.13}$$

On the other hand, differentiation of (1.10) with respect to s yields

$$\mathfrak{v}_i'\, \mathfrak{v}_k + \mathfrak{v}_i\, \mathfrak{v}_k' = 0$$

and hence, by (1.13),

$$a_{ik} + a_{ki} = 0. \tag{1.14}$$

Thus first of all, $a_{ii} = 0$. Further, (1.11) and (1.8) yield $a_{12} = \kappa$ and $a_{13} = 0$. Finally we define $\tau \underset{\text{def}}{=} a_{23}$ and we have obtained for the system (1.12) the following:

$$\begin{aligned} \mathfrak{v}_1' &= \phantom{-\kappa\mathfrak{v}_1} \quad \kappa\mathfrak{v}_2 \\ \mathfrak{v}_2' &= -\kappa\mathfrak{v}_1 \qquad\qquad + \tau\mathfrak{v}_3 \\ \mathfrak{v}_3' &= \phantom{-\kappa\mathfrak{v}_1} \quad -\tau\mathfrak{v}_2. \end{aligned} \tag{1.15}$$

Equations (1.15) dominate the entire theory of space curves. They were discovered and used independently by F. Frenet (1847, published 1852) and J. A. Serret (1851) and are referred to as the *Formulas of Frenet*.

1.6. Geometric Significance of Curvature and Torsion

Let us take a further look at the vectors $\mathfrak{v}_i$ and at the quantities κ and τ from the geometric point of view. The number τ is called the *torsion* (or *twisting number*) of the curve at the point $\mathfrak{x}(s)$. If $\tau(s) \equiv 0$, then it follows from the third of Frenet's formulas that $\mathfrak{v}_3(s) = \mathfrak{c}$, where $\mathfrak{c}$ is a constant vector. Since $\mathfrak{v}_1\mathfrak{v}_3 = 0$, it then follows that $\mathfrak{x}'\mathfrak{c} = 0$, hence $\mathfrak{x}\mathfrak{c} = \text{const.}$, so that $(\mathfrak{x}(s) - \mathfrak{x}_0)\mathfrak{c} = 0$ for a suitable vector $\mathfrak{x}_0$. Therefore the curve lies entirely in the plane through $\mathfrak{x}_0$ that is perpendicular to the vector $\mathfrak{c}$, and we have established the fact that *if* $\tau(s) \equiv 0$ *for a curve, then the curve is a plane curve.* The converse also holds, as can be verified without further calculation. For if $\mathfrak{x}(s)$ is a plane curve, then $\mathfrak{x}(s) = \mathfrak{x}_0 + a_1(s)\mathfrak{e}_1 + a_2(s)\mathfrak{e}_2$ with suitable vectors $\mathfrak{x}_0$, $\mathfrak{e}_1$, $\mathfrak{e}_2$ that are independent of s, and hence the four vectors

$$\mathfrak{v}_1, \quad \mathfrak{v}_2, \quad \mathfrak{v}_1', \quad \mathfrak{v}_2'$$

all lie in the plane spanned by $\mathfrak{e}_1$ and $\mathfrak{e}_2$, while the vector $\mathfrak{v}_3$ is perpendicular to this plane. Thus Frenet's second formula implies that $\tau \equiv 0$ in this case. We have shown, therefore, that *a curve* $\mathfrak{x}(s)$ *is a plane curve if, and only if, its torsion* $\tau(s)$ *vanishes identically.* Thus the torsion τ may be taken as a measure of how much a curve is twisted, that is to say, deviates from a plane course, just as the curvature κ measures deviation from a straight-line course. To follow this up further, let us investigate the behavior of the curve near (i.e., in the neighborhood of) a fixed point, at $s = s_0$. By Taylor's formula with remainder,

$$\mathfrak{x}(s_0 + \Delta s) = \mathfrak{x}(s_0) + \mathfrak{x}'(s_0)\,\Delta s + \tfrac{1}{2}\mathfrak{x}''(s_0)(\Delta s)^2 + \tfrac{1}{6}\mathfrak{x}'''(s_0)(\Delta s)^3 + \mathfrak{r}_4(\Delta s)$$

with $\lim_{\Delta s \to 0} \mathfrak{r}_4(\Delta s)/(\Delta s)^3 = 0$. For the derivatives we can substitute from Frenet's formulas, also using the fact that

$$\mathfrak{x}''' = (\mathfrak{x}'')' = (\kappa\mathfrak{v}_2)' = \kappa'\,\mathfrak{v}_2 + \kappa(-\,\kappa\mathfrak{v}_1 + \tau\mathfrak{v}_3).$$

Arranging, in Taylor's formula, according to multiples of the vectors $\mathfrak{v}_i = \mathfrak{v}_i(s_0)$, we thus obtain

$$\mathfrak{x}(s_0 + \Delta s) - \mathfrak{x}(s_0) = \mathfrak{v}_1 \Delta s\left(1 - \frac{\kappa^2}{6}\Delta s^2\right) \tag{1.16}$$

$$+ \mathfrak{v}_2 \Delta s^2\left(\frac{\kappa}{2} + \frac{\kappa'}{6}\Delta s\right) + \mathfrak{v}_3 \Delta s^3\,\frac{\kappa\tau}{6} + \mathfrak{r}_4.$$

This *canonical expansion* shows that the first-order approximation to the curve (i.e., an approximation that neglects terms in Δs of higher than the first degree) follows the tangent line to the curve, while the second approximation lies in the plane spanned by $\mathfrak{v}_1$ and $\mathfrak{v}_2$, the so-called *osculating plane.* The curvature κ is the coefficient of the second-degree term and measures the deviation from the tangent to the curve, while the torsion τ enters into the third-degree terms, which account for the deviation from the osculating plane. If we choose the coordinate axes for space in such a way that the origin is at $\mathfrak{x}(s_0)$ and that the unit vectors along the x, y, z-axes are $\mathfrak{v}_1(s_0)$, $\mathfrak{v}_2(s_0)$, $\mathfrak{v}_3(s_0)$ respectively, then the canonical expansion yields, neglecting terms of degree higher than the third in Δs:

$$x = \Delta s\left(1 - \frac{\kappa^2}{6}\Delta s^2\right),$$

$$y = \Delta s^2\left(\frac{\kappa}{2} + \frac{\kappa'}{6}\Delta s\right), \tag{1.17}$$

$$z = \Delta s^3\,\frac{\kappa\tau}{6}.$$

The three coordinate planes in the $\mathfrak{v}_1$, $\mathfrak{v}_2$, $\mathfrak{v}_3$-coordinate system are the $\mathfrak{v}_1$, $\mathfrak{v}_2$-plane, already introduced as the osculating plane; the $\mathfrak{v}_2$, $\mathfrak{v}_3$-plane,

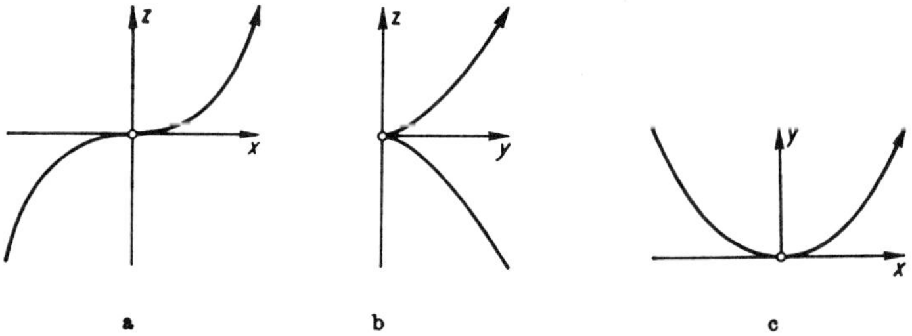

FIG. 4. Projections in the planes of the trihedron. (a) Rectifying plane. (b) Normal plane. (c) Osculating plane.

called the *normal plane* (it is perpendicular to the tangent line), and the $\mathfrak{v}_3$, $\mathfrak{v}_1$-plane, called the *rectifying plane.* To obtain the projections of the curve into these planes, we try to eliminate the parameter Δs from each of the three pairs of equations in (1.17), thus:

$$y = x^2\,\frac{\kappa}{2} + \Delta s^3(\ldots), \qquad \text{projection in the osculating plane;}$$

$$z^2 = y^3\,\frac{2\tau^2}{9\kappa} + \Delta s^7(\ldots), \qquad \text{projection in the normal plane;}$$

$$z = x^3\,\frac{\kappa\tau}{6} + \Delta s^4(\ldots), \qquad \text{projection in the rectifying plane.}$$

If $\kappa > 0$ and $\tau \neq 0$, the three projections behave at the point $\mathfrak{x}(s_0)$ — that is, for $\Delta s = 0$ — like a quadratic parabola, a semicubical (or Neil's) parabola, and a cubical parabola, respectively (see Fig. 4).

1.7. Osculating Circle and Osculating Sphere

A further geometric interpretation of the curvature κ is suggested by the formula for the projection of the curve in the osculating plane. The approximating parabola in the osculating plane has at its vertex an osculating circle whose radius (the radius of curvature) is $1/\kappa$. Let us see if it makes sense to call this number (the reciprocal curvature) the radius of curvature of the curve itself. We shall try to determine a sphere, the *osculating sphere*, which, at the given point of the curve, makes contact of as high an order as possible with the curve. The equation

$$(\mathfrak{x} - \mathfrak{m})^2 - r^2 = 0$$

represents a sphere with center at $\mathfrak{m}$ and radius r (assumed > 0). In order for this sphere to have the closest possible contact at $s = s_0$ with the curve $\mathfrak{x}(s)$, the function

$$f(s) = (\mathfrak{x}(s) - \mathfrak{m})^2 \qquad (1.18)$$

must equal r^2 at $s = s_0$ and must have as many vanishing derivatives there as possible. Using the Frenet formulas, we obtain

$$(') \qquad f'(s) = 2(\mathfrak{x} - \mathfrak{m})\mathfrak{v}_1$$

$$('') \qquad f''(s) = 2\kappa(\mathfrak{x} - \mathfrak{m})\mathfrak{v}_2 + 2$$

$$(''') \qquad f'''(s) = 2(\mathfrak{x} - \mu)(-\kappa^2\mathfrak{v}_1 + \kappa'\mathfrak{v}_2 + \kappa\tau\mathfrak{v}_3).$$

The conditions for contact of the second order are $f(s_0) = r^2$, $f'(s_0) = f''(s_0) = 0$. If we set

$$\mathfrak{x} - \mathfrak{m} = a_1\mathfrak{v}_1 + a_2\mathfrak{v}_2 + a_3\mathfrak{v}_3,$$

these conditions imply that $a_1 = (\mathfrak{x} - \mathfrak{m})\mathfrak{v}_1 = 0$ and $a_2 = (\mathfrak{x} - \mathfrak{m})\mathfrak{v}_2 = -1/\kappa$. Hence if $\kappa > 0$, we have

$$\mathfrak{m} = \mathfrak{x}(s_0) + \frac{1}{\kappa}\mathfrak{v}_2 - a_3\mathfrak{v}_3, \qquad (1.19)$$

where a_3 is arbitrary. Thus the straight line that is perpendicular to the osculating plane at the point $\mathfrak{x}(s_0) + (1/\kappa)\mathfrak{v}_2$ is the locus of the centers of all those spheres that have contact of at least second order with the curve at $\mathfrak{x}(s_0)$. This straight line is called the *polar axis* of the curve at s_0. The spheres just mentioned all contain the circle, lying in the osculating plane, with center at $\mathfrak{x}(s_0) + (1/\kappa)\mathfrak{v}_2$ and radius $1/\kappa$ and making contact of at least second order with the curve. This circle is called the *osculating circle* of the curve at s_0 and its center is the *center of curvature* of the curve at s_0 (see Fig. 5).

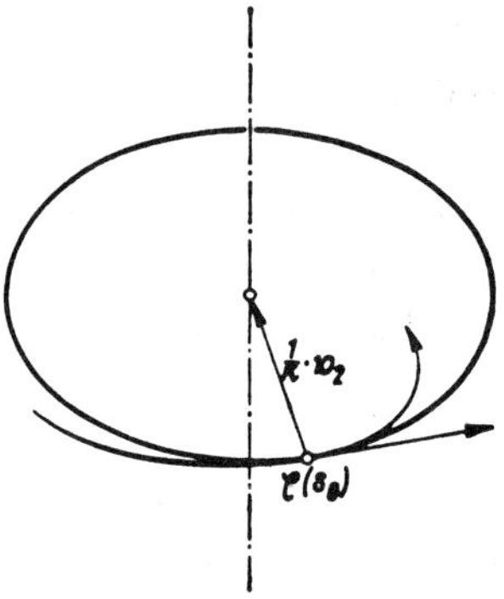

FIG. 5. Osculating circle and polar axis.

Next, if contact of at least third order is required, then the condition $f'''(s_0) = 0$ must be added to the foregoing, which implies, by (''') and (1.19), that

$$a_3 = \frac{\kappa'}{\tau\kappa^2}.$$

The unique sphere thus determined makes contact of (at least) third order with the curve at the point $\mathfrak{x}(s_0)$. It is called the *osculating sphere*; it center is

$$\mathfrak{m}_s = \mathfrak{x}(s_0) + \frac{1}{\kappa}\mathfrak{v}_2 - \frac{\kappa'}{\tau\kappa^2}\mathfrak{v}_3. \tag{1.20}$$

Exercise 1.7.1. Using $\tau = \mathfrak{v}_2'\mathfrak{v}_3$, *derive* the formula $\tau = \mathfrak{x}'\,\mathfrak{x}''\,\mathfrak{x}'''/(\mathfrak{x}'')^2$.

Exercise 1.7.2. *Prove* that each of the two following conditions characterizes straight lines: (a) All tangents pass through a fixed point. (b) All tangents are parallel to each other. [*Hint.* Use (1.15).] ← bad hint

Exercise 1.7.3. *Prove* that each of the two following conditions characterizes plane curves: (a) All osculating planes pass through a fixed point. (b) All osculating planes are parallel to each other. [*Hint.* Use (1.15).]

Exercise 1.7.4. If the initial points of all the unit tangent vectors $\mathfrak{v}_1$ are shifted to the origin, their new end points trace out a curve $\mathfrak{v}_1(s)$ on the unit sphere. This curve is called the *spherical indicatrix* (or *spherical tangent image*) of the curve $\mathfrak{x}(s)$. If $\tilde{s}$ denotes arc length on the spherical indicatrix, *show* that (see Fig. 6)

$$\kappa(s) = \frac{d\tilde{s}}{ds}.$$

Exercise 1.7.5. In analogy to Exercise 1.7.4, investigate the spherical binormal image and find a connection between its arc length and the torsion $\tau(s)$.

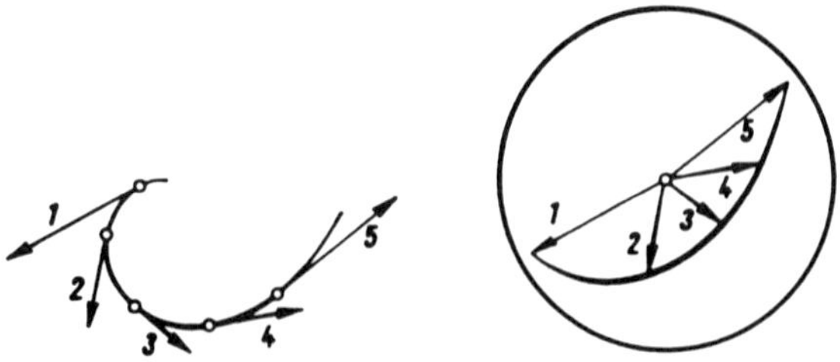

FIG. 6. Spherical indicatrix.

Exercise 1.7.6. *The Evolute of a Plane Curve* $\mathfrak{x}(s)$. Let $\mathfrak{x}(s)$ be a plane curve. Then the trace of the curve

$$\mathfrak{x}^*(s) = \mathfrak{x}(s) + \frac{1}{\kappa}\mathfrak{v}_2(s)$$

is the locus of the centers of the osculating circles. This curve $\mathfrak{x}^*(s)$ is called the *evolute* of the plane curve $\mathfrak{x}(s)$. *Show*: (a) The tangent to the evolute is in the direction of the principal normal of $\mathfrak{x}(s)$. (b) The distance $|\mathfrak{x}^*(s) - \mathfrak{x}(s)|$ equals, to within an additive constant, the arc length s^* of the evolute $\mathfrak{x}^*$ (see Fig. 7). [On account of property (b), the curve $\mathfrak{x}(s)$ can be generated as a string *involute* of $\mathfrak{x}^*$ if the evolute $\mathfrak{x}^*$ is given; holding it taut at one of its points, a string is wound off the curve $\mathfrak{x}^*$, onto which it has previously been wound.]

Exercise 1.7.7. *Spherical Curves.* *Prove* that the nonplanar curves that lie on a spherical surface are characterized by

$$\frac{\tau}{\kappa} = \left(\frac{\kappa'}{\tau\kappa^2}\right)'.$$

[*Hint.* The sphere must be the osculating sphere, at each point of such a curve; spherical curves are characterized by $\mathfrak{m}_s' = 0$. Use (1.20) and (1.15).]

Exercise 1.7.8. *Prove*: The osculating sphere is of constant radius if and only if the curve is either a spherical curve or a curve of constant curvature.

Exercise 1.7.9. *Prove*: The only (a) plane, (b) spherical curves of constant positive curvature are circles.

Exercise 1.7.10. *Prove* that the locus of the centers of curvature of a circular helix is another circular helix, with the same axis and of the same pitch. Prove also that circular helices have constant curvature and constant torsion.

Exercise 1.7.11. *The Rotation Vector of Darboux and Cesaro.* The Frenet formulas (1.15) can be interpreted kinematically, as follows: If a moving point traverses the curve $\mathfrak{x}(s)$ in such a way that s is the time parameter (hence, at uniform speed 1), then the moving trihedron $\mathfrak{v}_1$, $\mathfrak{v}_2$, $\mathfrak{v}_3$ moves in accordance with (1.15). This motion contains,

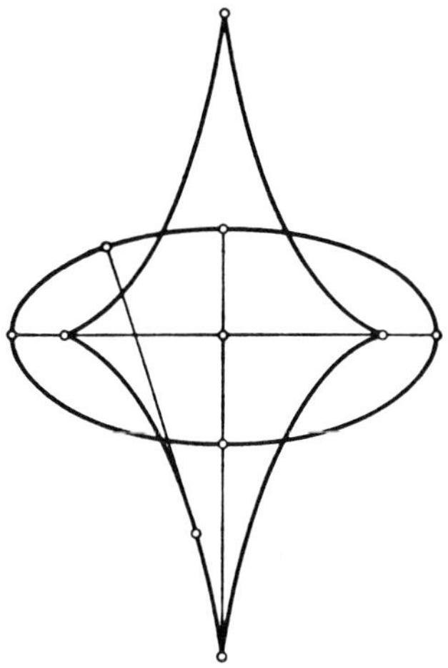

FIG. 7. Evolute (of the ellipse).

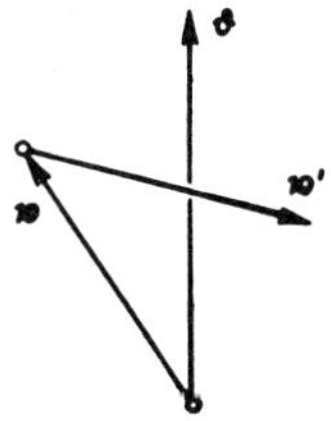

FIG. 8. Angular velocity vector $\mathfrak{d}$, and velocity $\mathfrak{v}' = \mathfrak{d} \times \mathfrak{v}$.

apart from an instantaneous translation, an instantaneous rotation with angular velocity (vector) $\mathfrak{d} = \tau\mathfrak{v}_1 + \kappa\mathfrak{v}_3$, since $\mathfrak{v}_i' = \mathfrak{d} \times \mathfrak{v}_i$ for $i = 1, 2, 3$. The direction of the vector $\mathfrak{d}$ is that of the instantaneous axis of rotation, and its length $\sqrt{\kappa^2 + \tau^2}$ is the (scalar) angular velocity (see Fig. 8).

§ 2. The Complete System of Invariants for Space Curves

2.1 Cylindrical Helices

All problems in the theory of curves tackled in the preceding section were dealt with successfully in terms of the curvature κ and the torsion τ, along with the vectors of the moving trihedron. We shall show in the present section that this was not due to any particular care in selecting the problems, but that it is indeed true that the two quantities κ and τ are sufficient for complete control over the differential geometry of space curves.

As a preliminary to documenting this statement, we deal first with a special class of space curves, the so-called *cylindrical* (or *general*) *helices*, also called *curves of constant slope*. The defining property of a cylindrical helix is that its curvature and torsion should maintain a constant ratio, thus: $\tau(s) = c \cdot \kappa(s)$. The following theorem establishes a second characteristic property:

Theorem 2.1.1. *A curve $\mathfrak{x}(s)$ is a cylindrical helix (i.e., has $\tau/\kappa \equiv$ const.) if, and only if, there is a unit vector $\mathfrak{e}_0$ for which $\mathfrak{e}_0\,\mathfrak{v}_1(s) \equiv$ const. Thus the cylindrical helices are precisely those curves whose tangents make a fixed angle with a fixed direction.*

Proof. First, assume that $\mathfrak{e}_0\,\mathfrak{x}' = \mathfrak{e}_0\,\mathfrak{v}_1 = \cos\varphi$, where φ is the fixed angle between the directions of the tangents and the direction of $\mathfrak{e}_0$. Differentiation yields $\mathfrak{e}_0\,\mathfrak{x}'' = 0$, hence $\mathfrak{e}_0\,\mathfrak{v}_2 = 0$. Hence, if we set $\mathfrak{e}_0 = \Sigma_{i=1}^{3} b_i\,\mathfrak{v}_i$, then $b_1 = \cos\varphi$, $b_2 = 0$, and, since $\mathfrak{e}_0$ is a unit vector, $b_3 = \sin\varphi$. Differentiating $\mathfrak{e}_0 = \mathfrak{v}_1 \cos\varphi + \mathfrak{v}_3 \sin\varphi$ and substituting from the first and third of Frenet's formulas, we obtain

$$0 = \kappa \cos\varphi - \tau \sin\varphi,$$

or $\tau = \kappa \cdot \cot\varphi$.

Conversely, if we assume that $\tau = c \cdot \kappa$ holds, then, setting $c = \cot\varphi$ and reversing the steps, we arrive at an equation $\mathfrak{e}_0 = \mathfrak{v}_1 \cos\varphi + \mathfrak{v}_3 \sin\varphi$, which in turn implies $\mathfrak{e}_0\mathfrak{v}_1 =$ const., q.e.d.

The cylindrical helices include as special cases all curves of torsion zero, that is, all plane curves, namely for $c = 0$. They also include the circular helices, namely for $\kappa =$ const., $c \neq 0$.

Next, let there be assigned a function $\kappa(s) > 0$ and a constant c and let us determine all curves with curvature $\kappa(s)$ and torsion $\tau(s) = c \cdot \kappa(s)$. By Frenet's formulas (1.15),

$$\mathfrak{v}_1' = \kappa(s)\mathfrak{v}_2, \qquad \mathfrak{v}_2' = -\,\kappa(s)\mathfrak{v}_1 + c\kappa(s)\mathfrak{v}_3, \qquad \mathfrak{v}_3' = -\,c\kappa(s)\mathfrak{v}_2.$$

If we introduce a new parameter t by means of

$$\frac{dt}{ds} = \kappa(s)$$

and if we denote derivatives with respect to t by dots, then Frenet's formulas turn into the following system of differential equations:

$$\dot{\mathfrak{v}}_1 = \mathfrak{v}_2, \qquad \dot{\mathfrak{v}}_2 = -\,\mathfrak{v}_1 + c\mathfrak{v}_3, \qquad \dot{\mathfrak{v}}_3 = -\,c\mathfrak{v}_2. \tag{2.1}$$

Differentiating the second of these equations and substituting from the first and third, we obtain

$$\dddot{\mathfrak{v}}_1 + \omega^2 \dot{\mathfrak{v}}_1 = 0, \qquad \text{where} \qquad \omega^2 = 1 + c^2. \tag{2.2}$$

The general solution of the system (2.2) is

$$\dot{\mathfrak{v}}_1 = \mathfrak{a} \cos \omega t + \mathfrak{b} \sin \omega t \qquad (\mathfrak{a}, \mathfrak{b}\text{: constant vectors}). \tag{2.3}$$

Since $\mathfrak{x}'' = \mathfrak{v}_1' = \dot{\mathfrak{v}}_1 \kappa(s)$, we can now obtain the general solution $\mathfrak{x}(s)$ of our problem by two consecutive integrations of (2.3) with respect to s, keeping in mind also that t must be expressed in terms of s. Thus we obtain the following:

Theorem 2.1.2. *If $\mathfrak{x}(s)$ is a cylindrical helix for which $\tau(s) = c \cdot \kappa(s)$, then*

$$\mathfrak{x}(s) = \frac{\mathfrak{a}}{\omega} \int_{s_0}^{s} \sin \omega t(\sigma)\, d\sigma - \frac{\mathfrak{b}}{\omega} \int_{s_0}^{s} \cos \omega t(\sigma)\, d\sigma + \mathfrak{c}_1 s + \mathfrak{c}_0 \tag{2.4}$$

where

$$t(s) = \int_{s_0}^{s} \kappa(\sigma)\, d\sigma \qquad \text{and} \qquad \omega^2 = 1 + c^2.$$

This theorem only gives necessary conditions. In order for $\mathfrak{x}(s)$, as given by (2.4), to actually be a cylindrical helix with curvature and torsion as specified, certain conditions have to be imposed on the constants of integration $\mathfrak{a}, \mathfrak{b}, \mathfrak{c}_1, \mathfrak{c}_0$. Since

$$\mathfrak{x}'(s) = \mathfrak{v}_1 = \frac{\mathfrak{a}}{\omega} \sin \omega t - \frac{\mathfrak{b}}{\omega} \cos \omega t + \mathfrak{c}_1$$

and

$$\mathfrak{x}''(s) = \kappa \mathfrak{v}_2 = \kappa(\mathfrak{a} \cos \omega t + \mathfrak{b} \sin \omega t),$$

it follows (on account of the orthogonality of the unit vectors $\mathfrak{v}_1, \mathfrak{v}_2$ for all t) that $\mathfrak{a}^2 = \mathfrak{b}^2 = 1$, $\mathfrak{a}\mathfrak{b} = \mathfrak{a}\mathfrak{c}_1 = \mathfrak{b}\mathfrak{c}_1 = 0$, and $\mathfrak{c}_1{}^2 = c^2/(1 + c^2)$. These conditions prove to be sufficient. If, incidentally, the initial position of the moving trihedron is given, then the constants of integration can be determined explicitly: for $s_0 = 0$, we have $\mathfrak{c}_0 = \mathfrak{x}(s_0)$, $\mathfrak{a} = \mathfrak{v}_2(s_0)$, $\mathfrak{c}_1 - \mathfrak{b}/\omega = \mathfrak{v}_1(s_0)$. The cotangent of the angle between $\mathfrak{v}_1$ and $\mathfrak{c}_1$ turns out to be the constant c, which shows that $\mathfrak{c}_1$ is in the direction with which the curve makes a fixed angle.

2.2 *The Fundamental Theorem*

Our next task will be to show that a similar state of affairs obtains for general curves. To be sure, we cannot expect to succeed in expressing $\mathfrak{x}(s)$ explicitly in terms of assigned functions $\kappa(s)$, $\tau(s)$ and assigned initial conditions $\mathfrak{x}(0)$, $\mathfrak{v}_i(0)$, as we did in the case of cylindrical helices. Rather, we shall have to be satisfied with the following existence and uniqueness theorem.

Theorem 2.2.1. *Let J be an open interval that contains the point* 0. *Let the two functions $\tilde{\kappa}(s)$, $\tilde{\tau}(s)$ be defined in J, let $\tilde{\kappa}(s)$ be positive and have a continuous first derivative, let $\tilde{\tau}(s)$ be continuous. Further, let $\mathfrak{c}_0$ be any vector and let the three vectors $\mathfrak{c}_i$ ($i = 1, 2, 3$) form an orthonormal trihedron. Then there exists one and only one curve $\mathfrak{x}(s)$ defined in J with the following properties:*

(a) *s is its arc length,*

(b) *$\mathfrak{x}(0) = \mathfrak{c}_0$ and $\mathfrak{v}_i(0) = \mathfrak{c}_i$, where the $\mathfrak{v}_i$ are the moving trihedron of $\mathfrak{x}(s)$,*

(c) *the curvature $\kappa(s)$ and the torsion $\tau(s)$ of the curve $\mathfrak{x}(s)$ satisfy*

$$\kappa(s) = \tilde{\kappa}(s), \qquad \tau(s) = \tilde{\tau}(s).$$

Proof. Following the proof of Theorem 2.1.2 as our model, we will first find a solution $\mathfrak{v}_i(s)$ of the system of Frenet's formulas, and we will then show that $\mathfrak{x}(s) = \int_0^s \mathfrak{v}_1(\sigma)\, d\sigma + \mathfrak{c}_0$ is a curve that has the required properties. According to the existence and uniqueness theorem for systems of linear ordinary differential equations (see Appendix II.1), the system of nine equations for nine unknown functions

$$\tilde{\mathfrak{v}}_i' = \sum_{k=1}^{3} a_{ik}\, \tilde{\mathfrak{v}}_k, \qquad (a_{ik}) = \begin{pmatrix} 0 & \tilde{\kappa} & 0 \\ -\tilde{\kappa} & 0 & \tilde{\tau} \\ 0 & -\tilde{\tau} & 0 \end{pmatrix} \tag{2.5}$$

has exactly one system of solutions $\tilde{\mathfrak{v}}_i(s)$ satisfying $\tilde{\mathfrak{v}}_i(0) = \mathfrak{c}_i$. Thus if there exists a curve $\mathfrak{x}(s)$ having the required properties, it is of necessity given by

$$\mathfrak{x}(s) = \mathfrak{c}_0 + \int_0^s \tilde{\mathfrak{v}}_1(\sigma)\, d\sigma. \tag{2.6}$$

This proves the uniqueness part of the theorem. The proof that $\mathfrak{x}(s)$ as given by (2.6) actually has the required properties will be given in several steps.

1. The $\mathfrak{x}(s)$ of (2.6) has continuous derivatives up to (and including) the third order. For, we have $\mathfrak{x}'(s) = \tilde{\mathfrak{v}}_1(s)$, and $\tilde{\mathfrak{v}}_1(s)$ is, by (2.5), differentiable, with derivative

$$\tilde{\mathfrak{v}}_1'(s) = \tilde{\kappa}(s)\,\tilde{\mathfrak{v}}_2(s);$$

this function, in turn, is itself differentiable, $\tilde{\kappa}(s)$ being (continuously) differentiable by assumption and $\tilde{\mathfrak{v}}_2(s)$ being so by virtue of (2.5).

2. For all s, we have $\tilde{\mathfrak{v}}_i\,\tilde{\mathfrak{v}}_k = \delta_{ik}$ and $\tilde{\mathfrak{v}}_1\,\tilde{\mathfrak{v}}_2\,\tilde{\mathfrak{v}}_3 = +1$. For, we have

$$(\tilde{\mathfrak{v}}_i\,\tilde{\mathfrak{v}}_k)' = \tilde{\mathfrak{v}}_i'\,\tilde{\mathfrak{v}}_k + \tilde{\mathfrak{v}}_i\,\tilde{\mathfrak{v}}_k' = \sum_{j=1}^{3} a_{ij}\,\tilde{\mathfrak{v}}_j\,\tilde{\mathfrak{v}}_k + \sum_{j=1}^{3} a_{kj}\,\tilde{\mathfrak{v}}_j\,\tilde{\mathfrak{v}}_i.$$

The theorem on systems of differential equations referred to above implies that the system

$$b_{ik}' = \sum_{j=1}^{3} (a_{ij}\,b_{jk} + a_{kj}\,b_{ji}),$$

with $b_{ik} = b_{ki}$ and the initial conditions $b_{ik}(0) = \delta_{ik}$, has exactly one solution. Since both $b_{ik}(s) = \delta_{ik}$ and $b_{ik}(s) = \tilde{\mathfrak{v}}_i\,\tilde{\mathfrak{v}}_k$ are solutions, the first part of the assertion is proved. The second part $(\tilde{\mathfrak{v}}_1\,\tilde{\mathfrak{v}}_2\,\tilde{\mathfrak{v}}_3 = 1)$ can be proved similarly.

3. s is the arc length of $\mathfrak{x}(s)$. For, by 2 above we have $\mathfrak{x}'^2 = \tilde{\mathfrak{v}}_1^{\,2} = 1$.

4. The $\tilde{\mathfrak{v}}_i(s)$ $(i = 1, 2, 3)$ are the vectors of the moving trihedron for $\mathfrak{x}(s)$. For, to begin with, the moving trihedron $\mathfrak{v}_i(s)$ $(i = 1, 2, 3)$ of $\mathfrak{x}(s)$ exists, since $\mathfrak{x}(s)$ has the requisite differentiability properties (by 1) and since $\mathfrak{x}''(s) = \mathfrak{v}_1'(s) = \tilde{\kappa}(s)\,\tilde{\mathfrak{v}}_2(s) \neq 0$ (by 2 and by $\tilde{\kappa}(s) > 0$). It follows that

$$\mathfrak{v}_1(s) = \mathfrak{x}'(s) = \tilde{\mathfrak{v}}_1(s).$$

From the definition (1.8), we have

$$\mathfrak{v}_2(s) = \frac{\mathfrak{x}''(s)}{|\mathfrak{x}''(s)|} = \frac{\tilde{\mathfrak{v}}_1'(s)}{\tilde{\kappa}(s)} = \tilde{\mathfrak{v}}_2(s),$$

using 1. Finally, from $\mathfrak{v}_1\,\mathfrak{v}_2\,\mathfrak{v}_3 = \tilde{\mathfrak{v}}_1\,\tilde{\mathfrak{v}}_2\,\tilde{\mathfrak{v}}_3 = 1$ and $\mathfrak{v}_i\,\mathfrak{v}_k = \tilde{\mathfrak{v}}_i\,\tilde{\mathfrak{v}}_k = \delta_{ik}$, it follows that $\mathfrak{v}_3 = \tilde{\mathfrak{v}}_3$.

5. $\tilde{\kappa}(s)$, $\tilde{\tau}(s)$ are the curvature and torsion, respectively, of $\mathfrak{x}(s)$. This follows from 4, if we compare the first and third of the equations in 1 with the first and third of Frenet's formulas. This completes the proof of Theorem 2.2.1.

2.3 *The Complete System of Invariants*

The important theorem proved in the preceding section can also be stated in a slighty different form. By means of a translation followed by a suitable rotation, we can first move any given point to the point $\mathfrak{c}_0$ and then bring any given orthonormal trihedron at that point into coincidence with the trihedron $\mathfrak{c}_i$ $(i = 1, 2, 3)$ at $\mathfrak{c}_0$. Since rigid motions leave curvature and torsion invariant, we have the following consequence of Theorem 2.2.1:

Fundamental Theorem of the Theory of Curves in Euclidean 3-Space. *If $\kappa(s)$, $\tau(s)$ are two functions satisfying the conditions of Theorem 2.2.1, then there exists one and, to within rigid motions, only one curve for which $\kappa(s)$, $\tau(s)$, and s are curvature, torsion, and arc length, respectively.*

The fundamental importance of this theorem lies in its implication that curvature and torsion constitute a complete system of invariants of a curve. By an invariant we here mean a functional[3] whose domain of definition includes all curves and which is invariant under rigid motions of (the trace of) a curve, i.e., whose value is the same for any two curves whose traces are congruent under a rigid motion. Let $I(s) = J[\mathfrak{x}(s)]$ be an invariant. By Theorem 2.2.1, a curve $\mathfrak{x}(s)$ is completely determined by its initial point $\mathfrak{x}(0)$, its moving trihedron at this point, and its curvature and torsion as functions of its arc length s. Now there is a rigid motion that maps the point $\mathfrak{x}(0)$ onto the origin of coordinates and that maps the moving trihedron at $\mathfrak{x}(0)$ onto the trihedron of unit vectors along the coordinate axes. This rigid motion leaves $\kappa(s)$ and $\tau(s)$ invariant, and it follows that $J[\mathfrak{x}(s)]$ is a function only of curvature and torsion. We can therefore state the following alternate version of the fundamental theorem:

Theorem 2.3.1. *Every invariant $J[\mathfrak{x}(s)]$ of a curve in Euclidean 3-space is a functional of curvature and torsion. These two functions, therefore, constitute a complete system of invariants for space curves.*

Since $\kappa(s)$ and $\tau(s)$ can be assigned independently of each other, these two functions even constitute a complete system of independent invariants. Therefore all geometric properties of a curve are completely determined by this pair of functions;

---

[3] The term *functional* is used for functions whose domain does not necessarily consist of numbers, but rather of functions or of other mathematical entities. For instance, the definite integral is a functional of functions (viz., of the integrands). The values (members of the range) of a functional need not be numbers either.

in other words, the geometry of the curve is completely known if its *natural* (or *intrinsic*) *equations* $\kappa = \kappa(s)$, $\tau = \tau(s)$ are given. At the same time, we can recognize a general method for solving problems in the differential geometry of space curves, in the following procedure: Write down all of the hypotheses, express all vectors occurring in them in terms of those of the moving trihedron, and replace the higher derivatives by using Frenet's formulas. In this manner the hypotheses can be transformed into relations involving only the vectors of the moving trihedron and the functions $\kappa(s)$, $\tau(s)$ (along with their derivatives). The solution can then, in general, be accomplished by comparing corresponding coefficients of the vectors of the moving trihedron and solving the resulting differential equations for $\kappa(s)$, $\tau(s)$. The reader should look back over the proofs of this chapter in the light of this general method and use the method for working out the exercises.

The reader is urged to make sure that he is quite clear about the concept of the complete system of invariants. The same conceptual structure, exemplified here in the simple case of the theory of curves, recurs in very similar form in the theory of surfaces and in other geometric systems.

Exercise 2.3.2. *Prove*: A curve for which $\kappa(s)$ = const. and $\tau(s)$ = const. is an ordinary (that is, circular) helix or one of its degenerate cases. (*Hint.* Since these curves are among the cylindrical helices, Theorem 2.1.2 may be used.)

Exercise 2.3.3. *Determine* all plane curves ($\tau = 0$) with $\kappa = \kappa(s)$, in accordance with Theorem 2.1.2 and integrate the resulting formulas for the particular cases (a) κ = const. (circles) and (b) $1/\kappa = c_1 s + c_2$ (logarithmic spirals).

Exercise 2.3.4. *Determine* the complete system of invariants for (a) triangles in the Euclidean plane, (b) tetrahedra in space, (c) cylindrical helices in space, (d) plane curves. In each case, investigate the question of independence.

Exercise 2.3.5. *Prove*: The spherical indicatrix of a curve is a circle if and only if the curve is a cylindrical helix.

Exercise 2.3.6. Denoting the arc length of the spherical principal normal image $\mathfrak{v}_2(s)$ by s^*, *prove* that $\kappa = |ds^*/ds|$ is a necessary and sufficient condition for a curve to be a plane curve.

CHAPTER II

Local Differential Geometry of Surfaces

§ 3. Surfaces, and Curves on Surfaces

3.1 Surface Parameters

Following Gauss, one defines surfaces in Euclidean 3-space in terms of mappings, *from* a set whose elements are given by pairs (u^1, u^2) of real numbers (real parameters) *into* 3-space. Thus a plane can be "parametrized" by a Cartesian coordinate system or by a polar coordinate system; a sphere, by geographic coordinates. When considering surfaces in the large ("globally"), one will not always find it possible to set up a one-to-one correspondence from a region of the Cartesian (u^1, u^2)-plane to the set of all points of the surface. Thus, for instance, the geographic coordinate system for a sphere fails at the two poles. But, inasmuch as local differential geometry confines itself to the investigation of sufficiently small portions of surfaces, the following definition will be sufficient for the time being (until we shall have to extend it, in §7 and in §17):

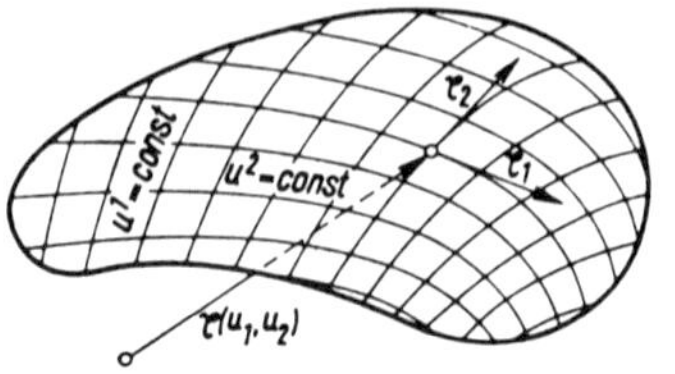

FIG. 9. Surface and parametric representation.

Definition. *A smooth portion of a surface (to be called a surface, for short) is given by*

$$\mathfrak{x} = \mathfrak{x}(u^i) = \mathfrak{x}(u^1, u^2) = \{x_1(u^1, u^2), x_2(u^1, u^2), x_3(u^1, u^2)\}, \tag{3.1}$$

where the pairs (u^i) range over a region G of the Cartesian (u^1, u^2)-plane while the functions x_i are of class[1] *C^3 (see Fig. 9). It is further assumed here that*

[1] We recall that a function is said to be of class C^k if all of its (ordinary, or partial) derivatives of orders $\leq k$ exist and are continuous. — For many parts of the theory of surfaces, less stringent differentiability conditions would be sufficient.

$$\mathfrak{x}_1 \times \mathfrak{x}_2 \neq 0, \qquad \textit{where} \qquad \mathfrak{x}_i \underset{\text{def}}{=} \frac{\partial \mathfrak{x}}{\partial u^i}. \tag{3.2}$$

We shall base our study of the local differential geometry of a surface upon the study of curves on the surface. A curve $\mathfrak{x} = \mathfrak{x}(t)$ is said to lie on (or in) the surface $\mathfrak{x} = \mathfrak{x}(u^i)$ if

$$\mathfrak{x}(t) = \mathfrak{x}(u^i(t)) \tag{3.3}$$

holds, where the functions $u^i(t)$ are of class C^3 and satisfy the condition $|\dot{u}^1| + |\dot{u}^2| \neq 0$ for all t in the domain of the curve $\mathfrak{x}(t)$.[2] The direction of the tangent to the surface curve (3.3) is that of the vector[3]

$$\dot{\mathfrak{x}} = \mathfrak{x}_1 \dot{u}^1 + \mathfrak{x}_2 \dot{u}^2 = \mathfrak{x}_i \dot{u}^i = \sum_{i=1}^{2} \mathfrak{x}_i \dot{u}^i. \tag{3.4}$$

If we construct all possible curves on the surface that pass through a fixed point $\mathfrak{x}_0 = \mathfrak{x}(\underset{0}{u^i})$ of the surface, we see from (3.4) that their tangent vectors point from $\mathfrak{x}_0$ into all possible directions within the plane spanned by the two vectors $\mathfrak{x}_k(\underset{0}{u^i})$. (The fact that these two vectors are linearly independent and therefore do span a plane, follows from the assumption $\mathfrak{x}_1 \times \mathfrak{x}_2 \neq 0$.) This plane is called the *tangent plane* of the surface at the point $\mathfrak{x}_0 = \mathfrak{x}(\underset{0}{u^i})$. The vector $\mathfrak{x}_1$ is the tangent vector at $\mathfrak{x}_0$ of the curve

$$u^1(t) = t, \qquad u^2(t) = \text{const.} = \underset{0}{u^2}.$$

A similar remark applies to the vector $\mathfrak{x}_2$. The curves $u^i = \text{const.}$ are called *parametric curves*; specifically, the curves $u^1 = \text{const.}$ are called the

---

[2] Here [as also earlier, in Eq. (1.4)] we are using one and the same letter with two different meanings: $\mathfrak{x}(t)$ and $\mathfrak{x}(u^i)$ are different functions. Such a procedure is often expedient and, in the present case, even logically unexceptionable, since the function symbols $\mathfrak{x}(\cdot)$ and $\mathfrak{x}(\cdot, \cdot)$ *are* distinct, because of the difference in their numbers of argument places.

[3] Here and in the sequel we use the very convenient *summation convention* introduced by A. Einstein. If an index occurs twice in a term, once as a superscript and once as a subscript [such as the index i in formula (3.4)], summation over that index is thereby indicated, specifically, in the theory of surfaces, from 1 through 2. The reader who is not used to this notation should at first write out all formulas in full; he will then recognize the usefulness of the convention soon enough.

u^2-curves (because only u^2 varies along such a curve) and, similarly, the curves $u^2 =$ const. are called the *u^1-curves*.

The following characterization of *geometric properties* parallels that given in the case of curves: *Geometric properties of a surface must be independent of*

(a) *the particular choice of a coordinate system in space,*

(b) *the particular choice of the parameters* (u^i) *for the surface.*

Since we write our formulas in vector notation, we are thereby assured of arriving at results that meet the first of these two invariance requirements. As to the second, i.e., as to invariance under (allowable) parameter transformations, this was accomplished in the theory of curves by the use of a particular invariant parameter. But in the theory of surfaces it will not do to always work in terms of any one particular system of parameters, as it will turn out that various different types of problems require the use of various different parametrizations. We shall therefore find it necessary to study parameter transformations more closely and to enter into separate discussions of invariance under such transformations [i.e., of the invariance property (b)] whenever a geometric property is being investigated. In a later chapter we shall take up the tensor calculus, whose notation is such that the question of invariance of a formula under parameter transformations can be decided immediately by inspection of the formula. It will also enable us to derive invariant formulas from invariant formulas. We begin with the definition of allowable parameter transformations.

Definition. *The functions*

$$u^i = u^i(\bar{u}^r) = u^i(\bar{u}^1, \bar{u}^2) \qquad (i = 1, 2) \tag{3.5}$$

are said to be an allowable parameter transformation of class C^k if they as well as their inverse functions $\bar{u}^r = \bar{u}^r(u^j)$ have continuous derivatives up to and including those of order k $(k \geqq 3)$, and if furthermore their functional determinant satisfies

$$\frac{\partial(u^i)}{\partial(\bar{u}^r)} \underset{\text{def}}{=} \begin{vmatrix} \dfrac{\partial u^1}{\partial \bar{u}^1} & \dfrac{\partial u^1}{\partial \bar{u}^2} \\ \dfrac{\partial u^2}{\partial \bar{u}^1} & \dfrac{\partial u^2}{\partial \bar{u}^2} \end{vmatrix} \neq 0. \tag{3.6}$$

The surface (3.1) can be related to the new parameters $\bar{u}^r$, as follows:

$$\mathfrak{x}(u^i) = \mathfrak{x}(u^i(\bar{u}^r)) = \bar{\mathfrak{x}}(\bar{u}^j).$$

If $k \geqq 3$, this parametrization of the surface satisfies the requirements of our definition of surface, since the functions $\bar{\mathfrak{x}}(\bar{u}^j)$ are of class C^3 and since the chain rule

$$\bar{\mathfrak{x}}_r \underset{\text{def}}{=} \frac{\partial \bar{\mathfrak{x}}}{\partial \bar{u}^r} = \mathfrak{x}_i \frac{\partial u^i}{\partial \bar{u}^r}, \tag{3.7}$$

together with condition (3.2), implies the corresponding inequality for the new parameters, on account of (3.6):

$$\bar{\mathfrak{x}}_1 \times \bar{\mathfrak{x}}_2 = \mathfrak{x}_i \frac{\partial u^i}{\partial \bar{u}^1} \times \mathfrak{x}_j \frac{\partial u^j}{\partial \bar{u}^2} = \mathfrak{x}_1 \times \mathfrak{x}_2 \cdot \frac{\partial(u^i)}{\partial(\bar{u}^k)} \neq 0. \tag{3.8}$$

3.2 *The Tangent Plane and the Moving Trihedron*

The first invariant of the surface that we shall study is the tangent plane. It consists of all vectors

$$\mathfrak{t} = \mathfrak{x}(\underset{0}{u^i}) + \mathfrak{x}_j(\underset{0}{u^i})\dot{u}^j,$$

and since

$$\dot{u}^j - \frac{\partial u^j}{\partial \bar{u}^r}\dot{\bar{u}}^r \qquad \text{and} \qquad \mathfrak{x}_j = \bar{\mathfrak{x}}_i \frac{\partial \bar{u}^i}{\partial u^j} \tag{3.9}$$

and

$$\frac{\partial \bar{u}^i}{\partial u^j}\frac{\partial u^j}{\partial \bar{u}^r} = \delta_r^i \underset{\text{def}}{=} \begin{cases} 0 & \text{for} \quad i \neq r \\ 1 & \phantom{\text{for}} \quad i = r \end{cases} \tag{3.10}$$

hold, we have

$$\mathfrak{t} = \mathfrak{x}_0 + \bar{\mathfrak{x}}_i \delta_r^i \dot{\bar{u}}^r = \mathfrak{x}_0 + \bar{\mathfrak{x}}_i \dot{\bar{u}}^i,$$

so that the tangent plane is given by the same vectors in a new parameter system as in the original one.

By (3.2), we can form the unit vector

$$\mathfrak{N} \underset{\text{def}}{=} \frac{\mathfrak{x}_1 \times \mathfrak{x}_2}{|\mathfrak{x}_1 \times \mathfrak{x}_2|}, \tag{3.11}$$

which is called the ***normal vector*** to the surface at the point in question, since it is perpendicular to all tangent vectors (3.4) (see Fig. 10). Under a parameter transformation, only the sign of the normal vector can change (*Exercise:* Prove this!), so that its direction is an invariant of the surface.

The trihedron $\mathfrak{x}_1$, $\mathfrak{x}_2$, $\mathfrak{N}$ plays an important role in the theory of surfaces, similar to that of the moving trihedron in the theory of curves. The former, however, is not orthonormal; we merely have

$$\mathfrak{N}^2 = 1, \qquad \mathfrak{N}\mathfrak{x}_i = 0.$$

Furthermore, this trihedron — unlike the moving trihedron of a space curve — is not invariant under parameter transformations.

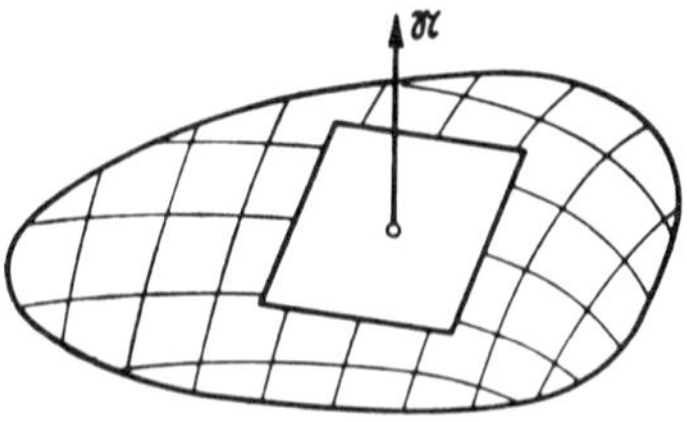

FIG. 10. Tangent plane and normal vector.

3.3 *Parametrizations Corresponding to Assigned Parametric Curves*

In the sequel we shall more than once have to arrange for the introduction of new parameters which are such that the curves of two given families become the new parametric curves. Because of its importance, we deal with this problem in two general theorems.

Theorem A. *Let*

$$A^{i}_{k}(u^j)\, du^k = 0, \qquad i = 1, 2, \qquad \det(A^{i}{}_{k}) \neq 0 \tag{3.12}$$

represent two given families of curves on a surface and assume that the functions A^{i}_{k} *are of class* C^n. *Then there is an allowable parameter transformation* $\bar{u}^i = \bar{u}^i(u^j)$ *of class* C^n *which is such that the new parametric lines are the assigned curves* (3.12).

Proof. The $\bar{u}^i$ can be determined from $\partial \bar{u}_i/\partial u_k = \lambda^i(u) A^{i}_{k}(u^j)$. The functional determinant does not vanish, by assumption. The existence of the integrating factors λ^i is known from the theory of differential equations.

Theorem B. *Let*

$$B_{ik}(u^j)\, du^i\, du^k = 0, \qquad \text{with} \qquad B_{ik} = B_{ki}, \qquad \det(B_{ik}) < 0, \tag{3.13}$$

represent two assigned families of curves on a surface. (The determinant must be negative, since otherwise we would not have two real, distinct families of curves.) *Assume that the functions* B_{ik} *are of class* C^n. *Then there is an allowable parameter transformation* $\bar{u}^i = \bar{u}^i(u^j)$ *of class* C^n, *defined in a sufficiently small region, such that the curves* (3.13) *are the new parametric curves.*

Proof: If $B_{11} = B_{22} = 0$ throughout an open portion of the surface, then (3.13) implies that $B_{12}\, du^1\, du^2 = 0$, so that we can set $\bar{u}^i = u^i$. Otherwise, one of the functions B_{ii}, say B_{11}, does not vanish in some sufficiently small open portion of the surface. By multiplying out, we verify that

$$B_{11} \cdot B_{ik}\, du^i\, du^k$$
$$= (B_{11}\, du^1 + (B_{12} + \sqrt{-\det B_{ik}})\, du^2) \cdot (B_{11}\, du^1 + (B_{12} - \sqrt{-\det B_{ik}})\, du^2),$$

so that we can now refer back to Theorem A by setting

$$A_1^1 = B_{11}, \qquad A_2^1 = B_{12} + \sqrt{-\det B_{ik}},$$
$$A_1^2 = B_{11}, \qquad A_2^2 = B_{12} - \sqrt{-\det B_{ik}},$$

which implies that

$$\det A_k^i = -2B_{11}\sqrt{-\det(B_{ik})} \neq 0.$$

Exercise 3.3.1. *A Parametric Representation for Surfaces of Revolution.* A plane curve generates a surface of revolution if its plane is rotated about a straight line in the plane. We may take this straight line to be the x_3-axis and we introduce r and x_3 as Cartesian coordinates in the plane. Let the curve be given by

$$r = r(t) \geqq 0, \qquad x_3 = x_3(t)$$

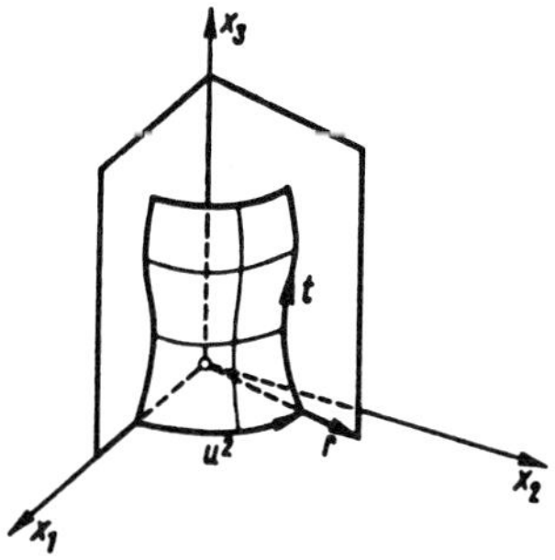

FIG. 11. Generating a surface of revolution.

(see Fig. 11). If we introduce the parameters $u^1 = t$, $u^2 =$ angle of rotation of the plane about the x_3-axis, then, for the points of the surface of revolution, we obtain the representation

$$\mathfrak{x} = \mathfrak{x}(u^i) = \{r(u^1)\cos u^2,\, r(u^1)\sin u^2,\, x_3(u^1)\}.$$

Show that $|\mathfrak{x}_1 \times \mathfrak{x}_2| = r\sqrt{\dot{x}_3^2 + \dot{r}^2}$, which implies that condition (3.2) is satisfied for all points other than those on the axis of rotation ($r = 0$). [The radical is $\neq 0$ for allowable curve parameters t, on account of (1.1).]

Show also that $\dot{x}_3^2 + \dot{r}^2 = 1$ if t is the arc length of the curve, and compute the normal vector.

The u^1-curves are called *meridians*, the u^2-curves are called *circles of latitude*.

Exercise 3.3.2. Let a surface be given by $x_3 = f(x_1, x_2)$. Setting $u^1 = x_1$, $u^2 = x_2$, we obtain a parametrization of this surface as follows:

$$\mathfrak{x} = \mathfrak{x}(u^i) = \{u^1, u^2, f(u^1, u^2)\}.$$

Determine under what conditions the requirement (3.2) is satisfied, and compute the normal vector.

3.4 Arc Length and First Fundamental Form

We introduce some basic quantities associated with the surface, basing our study on geometric (invariant) quantities associated with the curves on the surface. In this section, we shall use the arc length of curves on the surface for studying the geometry of the surface itself, and in the next section we shall similarly use the curvature of these curves.

The arc length of a curve $\mathfrak{x}(t) = \mathfrak{x}(u^i(t))$ on the surface is given by

$$s(t) = \int^{t} \sqrt{\dot{\mathfrak{x}}^2}\, dt, \tag{3.14}$$

so that, since

$$\dot{\mathfrak{x}}^2 = \dot{\mathfrak{x}}\,\dot{\mathfrak{x}} = \mathfrak{x}_i\,\dot{u}^i\,\mathfrak{x}_k\,\dot{u}^k = \mathfrak{x}_i\,\mathfrak{x}_k\,\dot{u}^i\,\dot{u}^k, \tag{3.15}$$

we can calculate this arc length, provided we know the three quantities,

$$g_{ik} \underset{\text{def}}{=} \mathfrak{x}_i\,\mathfrak{x}_k, \qquad g_{ik} = g_{ki}. \tag{3.16}$$

The functions $g_{ik} = g_{ik}(u^j)$, introduced by Gauss, are called the *metric coefficients* (in the older literature, these are often denoted by the letters E, F, G), and the quadratic form

$$ds^2 = g_{ik}(u^j)\,du^i\,du^k \tag{3.17}$$

is called the *first* (*or metric*) *fundamental form* of the surface. One also refers to ds as the element of arc on the surface.

We have $g_{ii} = \mathfrak{x}_i^2 > 0$ and

$$g \underset{\text{def}}{=} \det(g_{ik}) = g_{11}\,g_{22} - g_{12}^2 = \mathfrak{x}_1^2\,\mathfrak{x}_2^2 - (\mathfrak{x}_1\,\mathfrak{x}_2)^2 = (\mathfrak{x}_1 \times \mathfrak{x}_2)^2 > 0. \tag{3.18}$$

The form (3.17) is positive-definite, a fact which already resides in its geometric significance.

The terminology "metric fundamental form" is explained by the fact that the g_{ik} can be used for measuring not only lengths on the surface, but also angles and areas on the surface. Specifically, if two curves on the surface are given by $u^i = u^i(t)$ and $u^i = u^{*i}(t^*)$, respectively, both passing through the point $\mathfrak{x}$ on the surface for $t = t^* = 0$, then the angle γ between their tangents at $\mathfrak{x}$ is given by

$$\cos\gamma = \frac{\dot{\mathfrak{x}}\,\dot{\mathfrak{x}}^*}{|\dot{\mathfrak{x}}|\cdot|\dot{\mathfrak{x}}^*|} = \frac{\mathfrak{x}_i\,\dot{u}^i\,\mathfrak{x}_k\,\dot{u}^{*k}}{\sqrt{\mathfrak{x}_i\,\mathfrak{x}_k\,\dot{u}^i\,\dot{u}^k\,\mathfrak{x}_l\,\mathfrak{x}_m\,\dot{u}^{*l}\,\dot{u}^{*m}}} = \frac{g_{ik}\,\dot{u}^i\,\dot{u}^{*k}}{\sqrt{g_{ik}\,\dot{u}^i\,\dot{u}^k\,g_{lm}\,\dot{u}^{*l}\,\dot{u}^{*m}}}. \tag{3.19}$$

The area of a portion of the surface is given by

$$\iint \mathfrak{N}\mathfrak{x}_1\,\mathfrak{x}_2\,du^1\,du^2. \tag{3.20}$$

We have

$$\mathfrak{N}\mathfrak{x}_1\,\mathfrak{x}_2 = \frac{(\mathfrak{x}_1 \times \mathfrak{x}_2)^2}{|\mathfrak{x}_1 \times \mathfrak{x}_2|} = +\sqrt{g}$$

on account of Eq. (3.18).

Thus, as shown in (3.19) and (3.20), we can express angles and areas, as well as lengths, in terms of the metric coefficients g_{ik}.

We now have to prove that the formulas we have obtained are invariant under transformations of the coordinate system in space as well as under parameter transformations. The first invariance property is assured by the use of vector notation in these formulas. As to the second invariance property, we first note that the quantities g_{ik} separately are certainly not invariant. Under a parameter transformation (3.5) we obtain, by (3.7),

$$\bar{g}_{ik} = \bar{\mathfrak{x}}_i\,\bar{\mathfrak{x}}_k = \mathfrak{x}_l\,\mathfrak{x}_m\,\frac{\partial u^l}{\partial \bar{u}^i}\,\frac{\partial u^m}{\partial \bar{u}^k} = g_{lm}\,\frac{\partial u^l}{\partial \bar{u}^i}\,\frac{\partial u^m}{\partial \bar{u}^k}. \tag{3.21}$$

From this, it follows that (3.15), and thus the arc length (3.14), are indeed invariant:

$$\dot{\mathfrak{x}}^2 = \bar{g}_{ik}\,\dot{\bar{u}}^i\,\dot{\bar{u}}^k = g_{lm}\,\frac{\partial u^l}{\partial \bar{u}^i}\,\frac{\partial u^m}{\partial \bar{u}^k}\,\frac{\partial \bar{u}^i}{\partial u^j}\,\frac{\partial \bar{u}^k}{\partial u^n}\,\dot{u}^j\,\dot{u}^n$$

$$= g_{lm}\,\delta_i^l\,\delta_k^m\,\dot{u}^i\,\dot{u}^k = g_{ik}\,\dot{u}^i\,\dot{u}^k = \dot{\mathfrak{x}}^2.$$

This result on the invariance of arc length was, of course, to be expected on geometric grounds.

Exercise 3.4.1. (a) *Prove* that the two families of parametric curves form an orthogonal net (i.e., that the curves of one family intersect those of the other at right angles) if and only if $g_{12} \equiv 0$. (b) *Prove* that a necessary and sufficient condition for u^1 to be the arc length for a u^1-curve is that $g_{11} = 1$ along this curve.

Exercise 3.4.2. *Prove* that the meridians and circles of latitude of a surface of revolution (see Exercise 3.3.1) form an orthogonal net.

Exercise 3.4.3. Let a surface be given as in Exercise 3.3.2. Calculate the g_{ik} and the determinant g. What form does the area formula (3.20) assume in this case?

Exercise 3.4.4. *Check* on the invariance of angles and areas under parameter transformations.

Exercise 3.4.5. The plane can be parametrized by Cartesian coordinates u^1, u^2 and it can also be parametrized by polar coordinates $\bar{u}^1 = r$, $\bar{u}^2 = \varphi$; these are related by the parameter transformation

$$u^1 = \bar{u}^1 \cos \bar{u}^2, \qquad u^2 = \bar{u}^1 \sin \bar{u}^2.$$

Prove that this is an allowable transformation for $r > 0$. In the u-coordinate system we have $g_{ik} = \delta_{ik}$; use (3.21) to derive the $\bar{g}_{ik}$.

Exercise 3.4.6. *Axiomatic Treatment of Area.* Formula (3.20) for area is justified insofar as it becomes the familiar area formula of the integral calculus if the parameters are the special parameters of Exercise 3.4.3 (or 3.3.2), as follows from Exercise 3.4.3. A more satisfactory justification can be given as follows: To ensure that the area of a portion of surface consisting of two subportions should equal the sum of the areas of these two parts (i.e., to ensure that area should be finitely additive), we try for the area A an expression of the form

$$A = \iint_{\mathfrak{U}} F \, du^1 \, du^2.$$

The function F here is subject to the following requirements:

(a) $\iint F \, du^1 \, du^2$ is an invariant under rigid motions,

(b) $\iint F \, du^1 \, du^2$ is invariant under parameter transformations that preserve orientation [i.e., parameter transformations $\bar{u}^i = \bar{u}^i(u^j)$ with $\partial(\bar{u}^i)/\partial(u^j) > 0$],

(c) the area of a plane square of side 1 equals unity,

(d) $F = F(\mathfrak{x}, \mathfrak{x}_1, \mathfrak{x}_2, u^1, u^2)$.

The only one of these conditions that may lack intuitive plausibility is (d). However, the dependence stipulated by (d) is the simplest dependence one could expect.

Show: $F = \sqrt{g}$ is the only function for which (a) through (d) are satisfied. [*Hint.* First, it is easy to check directly that $\sqrt{g}$ does have the required properties. Second, conversely, let F be a function for which properties (a) through (d) are satisfied. Then, by (a) and (b), the quotient

$$\iint_{\mathfrak{U}} F \, du^1 \, du^2 \Big/ \iint_{\mathfrak{U}} \sqrt{g} \, du^1 \, du^2$$

also has invariance properties (a) and (b) for arbitrary regions $\mathfrak{U}$ of the parameters, so that we see, by letting $\mathfrak{U} \to 0$, that $F/\sqrt{g}$ is invariant. Exploit this first for transformations $\bar{u}^i = u^i + c^i$ (c^i = const.), which leave $\mathfrak{x}$, $\mathfrak{x}_1$, $\mathfrak{x}_2$ invariant, thus recognizing that F cannot explicitly depend on the u^i. Thus $F = F(\mathfrak{x}, \mathfrak{x}_1, \mathfrak{x}_2)$. Now under a translation in space, $\mathfrak{x}$ varies but the $\mathfrak{x}_i$ do not; hence $F = F(\mathfrak{x}_1, \mathfrak{x}_2)$. Finally, be means of a suitable parameter transformation the $\mathfrak{x}_i$ can be transformed into any other pair of vectors. Under this transformation, $F/\sqrt{g}$ must remain invariant and must, therefore, be a constant, whose value is 1 on account of (c).]

3.5 The Curvature of Curves on the Surface

Having made use of arc length, we shall next exploit the curvature of curves on the surface in a similar way for studying the geometry of the surface. Let $\mathfrak{x}(s) = \mathfrak{x}(u^i(s))$ be a curve on the surface, parametrized by its arc length. Then

$$\mathfrak{x}' = \mathfrak{x}_i\, u^{i\prime} \tag{3.22}$$

and

$$\mathfrak{x}''(s) = \mathfrak{x}_i\, u^{i\prime\prime} + \mathfrak{x}_{ik}\, u^{i\prime}\, u^{k\prime}, \qquad \mathfrak{x}_{ik} \underset{\text{def}}{=} \frac{\partial^2 \mathfrak{x}}{\partial u^i\, \partial u^k}. \tag{3.23}$$

By the first of Frenet's formulas,

$$\mathfrak{x}''(s) = \mathfrak{v}_1'(s) = \kappa(s)\mathfrak{v}_2(s), \tag{3.24}$$

the vector (3.23) — the curvature vector of the curve — has length $\kappa(s)$. We now recall that in the theory of space curves we expressed all vectors in terms of the moving trihedron $\mathfrak{v}_1$, $\mathfrak{v}_2$, $\mathfrak{v}_3$. In the theory of surfaces, we shall call on the moving trihedron $\mathfrak{x}_1$, $\mathfrak{x}_2$, $\mathfrak{N}$ in the same way and we shall thus break down every vector into a tangential and a normal component. For the curvature vector $\mathfrak{x}''$ we set (see Fig. 12)

$$\mathfrak{x}''(s) = \kappa_n\, \mathfrak{N} + \kappa_g\, \mathfrak{s} \tag{3.25}$$

where $\mathfrak{s}$ is a unit tangent vector, so that $\mathfrak{s}^2 = 1$, $\mathfrak{s}\mathfrak{N} = 0$. Since $\mathfrak{x}'^2 = 1$ yields $\mathfrak{x}'\, \mathfrak{x}'' = 0$ upon differentiation (with respect to arc length), we obtain from (3.25), by forming scalar products with $\mathfrak{x}'$,

$$0 = \kappa_g\, \mathfrak{x}'\mathfrak{s},$$

from which it follows if we fix the sign of κ_g suitably, that

$$\mathfrak{s} = \mathfrak{N} \times \mathfrak{x}', \tag{3.26}$$

since $\mathfrak{s}$ is a unit vector that is perpendicular to the orthonormal set of vectors $\mathfrak{N}$, $\mathfrak{x}'$. [If $\kappa_g = 0$, we cannot draw the conclusion that $\mathfrak{x}'\mathfrak{s} = 0$, but in this case we have complete freedom in the choice of $\mathfrak{s}$ and we shall define $\mathfrak{s}$ by (3.26) in this case also.]

We have $\kappa^2 = \mathfrak{x}''^2 = \kappa_n^2 + \kappa_g^2$. The number κ_n, which is the magnitude of the normal component of the curvature vector, is called the *normal curvature*, and the magnitude κ_g of the tangential component is called the

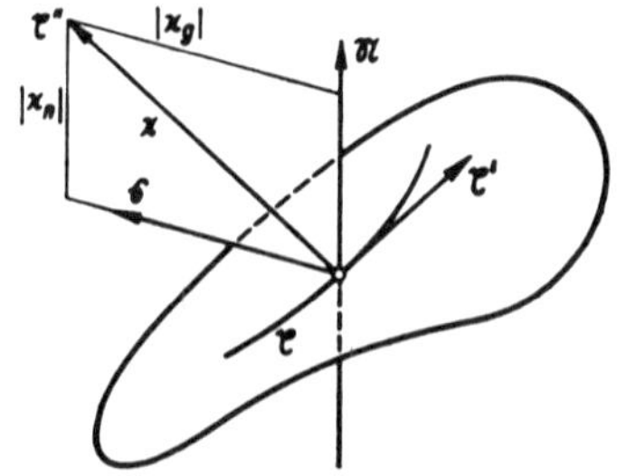

FIG. 12. Curvature vector of a curve on the surface.

geodesic curvature, of the curve on the surface. Toward calculating these quantities by means of (3.23), we first break down the vectors $\mathfrak{x}_{ik}$ into their tangential and normal components. We set

$$\mathfrak{x}_{ik} = \Gamma^r_{ik}\, \mathfrak{x}_r + L_{ik}\, \mathfrak{N}, \tag{3.27}$$

and, in these *formulas of Gauss*, we still have to determine the coefficients Γ^r_{ik}, L_{ik}. Forming the inner product of (3.27) with $\mathfrak{N}$, we obtain

$$\mathfrak{N}\mathfrak{x}_{ik} = L_{ik},$$

and, since differentiation with respect to u^k of $\mathfrak{N}\mathfrak{x}_i = 0$ yields

$$\mathfrak{N}\mathfrak{x}_{ik} + \mathfrak{N}_i\, \mathfrak{x}_k = 0, \qquad \mathfrak{N}_i \underset{\text{def}}{=} \frac{\partial \mathfrak{N}}{\partial u^i}, \tag{3.28}$$

we have

$$L_{ik} = \mathfrak{N}\mathfrak{x}_{ik} = -\,\mathfrak{N}_i\, \mathfrak{x}_k, \qquad L_{ik} = L_{ki}. \tag{3.29}$$

The three functions L_{ik} are the *coefficients of the second fundamental form* of the surface and we will study them in Section 5 below. We postpone for the moment the determination of the quantities Γ^r_{ik}.

If we use (3.27), (3.23) becomes

$$\mathfrak{x}'' = \mathfrak{x}_r(u''^r + \Gamma^r_{ik}\, u'^i\, u'^k) + \mathfrak{N} \cdot L_{ik}\, u'^i\, u'^k, \tag{3.30}$$

from which we obtain, on account of (3.25),

$$\kappa_n = L_{ik}\, u'^i\, u'^k \tag{3.31}$$

and

$$\kappa_g\, \mathfrak{s} = \mathfrak{x}_r(u''^r + \Gamma^r_{ik}\, u'^i\, u'^k). \tag{3.32}$$

The quantities L_{ik}, Γ^r_{ik} were introduced on a formal basis. A geometric meaning attaches to these quantities. It will become clear what the geometric significance of these numbers is, once we investigate more closely the geodesic and normal curvature, which we shall do in the two following chapters. It will turn out that these quantities are completely sufficient for the local study of a surface, in the same sense in which, according to the fundamental theorem of the theory of space curves, curvature and torsion suffice for the local study of space curves.

For this reason it turns out to be superfluous to study the torsion of curves in the surface as a tool for the study of the surface itself. The torsion would add nothing to what has been contributed by arc length and curvature and can be expressed in terms of the quantities introduced above.

Exercise 3.5.1. Let a meridian of a surface of revolution be parametrized by its arc length. *Prove* that

$$L_{11} = x_3'' r' - r'' x_3', \qquad L_{12} = 0, \qquad L_{22} = r x_3'$$

and use this to prove the following: The only surfaces of revolution for which $\det(L_{ik}) = 0$ are the circular cones (including the degenerate cases of circular cylinders and planes), that is, precisely those surfaces of revolution whose meridians are straight lines.

Exercise 3.5.2. *Prove*: Under a parameter transformation with positive functional determinant the L_{ik} transform in the same way as do the g_{ik} [see (3.21)].

3.6 Ruled Surfaces

Before continuing with a systematic development of the theory of surfaces, we shall investigate in some detail a special class of surfaces of particular importance, the so-called *ruled surfaces*. This will give us an opportunity to introduce the concept of *isometric mapping*, or *application*, of two surfaces. Ruled surfaces are defined as surfaces whose trace is the locus of a moving straight line in space. Through every point of such a surface, there must therefore pass a straight line all of whose points belong to the surface; such a straight line is called a *generator* of the ruled surface. If $\mathfrak{y}(u^1)$ is the trajectory of a point on the moving straight line and if the direction of the straight line is given by the unit vector $\mathfrak{z}(u^1)$, then the surface can be represented in the form

$$\mathfrak{x}(u^1, u^2) = \mathfrak{y}(u^1) + u^2\,\mathfrak{z}(u^1) \tag{3.33}$$

(see Fig. 13). In order for this to be a surface in the sense of the definition given in §3.1, we have to assume here that the functions $\mathfrak{y}$ and $\mathfrak{z}$ are of class C^3 and that

$$\mathfrak{x}_1 \times \mathfrak{x}_2 = (\mathfrak{y}' + u^2\,\mathfrak{z}') \times \mathfrak{z} \neq 0. \tag{3.34}$$

As examples of ruled surfaces, we cite the cones (where all generators pass through a fixed point) and the cylinders (where all generators are parallel

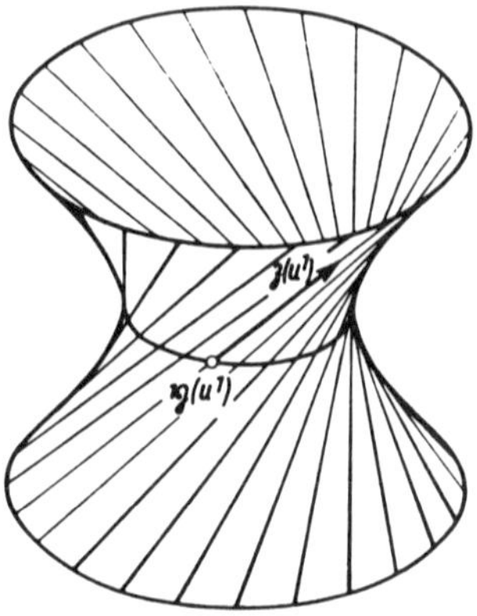

FIG. 13. Ruled surface, hyperboloid of one sheet.

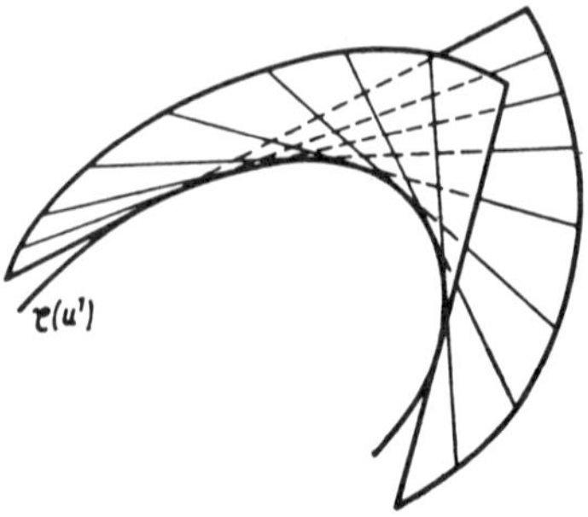

FIG. 14. Tangential developable (tangent surface).

to a fixed direction). Further examples of ruled surfaces are furnished by the *tangential developables* (or *tangent surfaces*), whose generators are the tangents of a fixed space curve $\mathfrak{x}(u^1)$, so that their points can be represented in the form

$$\mathfrak{x}(u^1, u^2) = \mathfrak{x}(u^1) + u^2\,\mathfrak{x}'(u^1) \tag{3.35}$$

(see Fig. 14). If u^1 is the arc length along the curve $\mathfrak{x}(u^1)$, then

$$\mathfrak{x}_1 \times \mathfrak{x}_2 = (\mathfrak{x}' + u^2\,\mathfrak{x}'') \times \mathfrak{x}' = -\,u^2\,\kappa\mathfrak{v}_3 \tag{3.36}$$

where κ is the curvature of the curve $\mathfrak{x}(u^1)$ and $\mathfrak{v}_3$ is its binormal. Hence, for a space curve of positive curvature, the tangents generate a surface, apart from points of the curve itself ($u^2 \neq 0$). The curve itself is called the *edge of regression* of its tangent surface; the tangent surface becomes singular at the points of its edge of regression, whence the term "edge."

For the normal vector to a tangent surface we obtain, from (3.36),

$$\mathfrak{N}(u^i) = \pm \mathfrak{v}_3(u^1).$$

This vector therefore is independent of u^2, which shows that the tangent planes at all points of a fixed generator coincide. A ruled surface with the property just described is called a *developable surface.*

Definition. *A ruled surface is called a developable surface if the tangent plane at any one of its points is also the tangent plane at all other points of the generator passing through this point.*

Exercise. Prove that not only the tangent surfaces, but also all cones and cylinders are developable surfaces.

Theorem 3.6.1. *A ruled surface* (3.33) *is developable if and only if* $\mathfrak{y}'\mathfrak{z}\mathfrak{z}' = 0$.

Proof. First, a developable surface is characterized by the condition $\mathfrak{N}_2 = 0$, since the normal vector is constant along every generator. Next, differentiation with respect to u^2 of $\mathfrak{N}^2 = 1$ yields $\mathfrak{N}\mathfrak{N}_2 = 0$, from which it follows that the conditions $\mathfrak{N}_2 = 0$ and $\mathfrak{N} \times \mathfrak{N}_2 = 0$ are equivalent. But, by (3.34) and (3.18), we have

$$\mathfrak{N} = \frac{1}{\sqrt{g}}(\mathfrak{y}' + u^2\,\mathfrak{z}') \times \mathfrak{z}$$

and

$$\mathfrak{N}_2 = \frac{1}{\sqrt{g}}\mathfrak{z}' \times \mathfrak{z} + \mathfrak{N}\sqrt{g}\,\frac{\partial}{\partial u^2}\left(\frac{1}{\sqrt{g}}\right);$$

therefore

$$\mathfrak{N} \times \mathfrak{N}_2 = \frac{1}{g}\cdot(\mathfrak{y}' \times \mathfrak{z}) \times (\mathfrak{z}' \times \mathfrak{z}) = -\frac{\mathfrak{z}}{g}(\mathfrak{y}'\mathfrak{z}\mathfrak{z}'),$$

so that a necessary and sufficient condition for $\mathfrak{N} \times \mathfrak{N}_2 = 0$ is that $\mathfrak{y}'\mathfrak{z}\mathfrak{z}' = 0$, which proves the theorem.

3.7 Developable Surfaces

The next theorem gives an explicit determination of all developable surfaces.

Theorem 3.7.1. *The totality of developable surfaces is the same as the totality of tangential developables, cones, and cylinders.*

Proof. We already know that the three last-named types of surfaces are developable surfaces. Conversely, let there be given a ruled surface that is developable, so that the condition given in Theorem 3.6.1 applies. Without loss of generality, we may assume that the directrix $\mathfrak{y}(u^1)$ intersects every generator at right angles and that u^1 is the arc length of $\mathfrak{y}$, so that $\mathfrak{y}'\,\mathfrak{z} = 0$. Now consider any curve on the surface determined by, say, $u^2 = v(u^1)$, i.e., $\mathfrak{x}(u^1) = \mathfrak{x}(u^1, v(u^1))$. By (3.33), we have

$$\mathfrak{x}' = \frac{d\mathfrak{x}(u^1)}{du^1} = \mathfrak{y}' + \mathfrak{v}\mathfrak{z}' + \mathfrak{v}'\,\mathfrak{z}.$$

Since $\mathfrak{y}', \mathfrak{z}, \mathfrak{N}$ are three mutually perpendicular unit vectors, we have

$$\mathfrak{z}' = a\mathfrak{y}' + b\mathfrak{z} + c\mathfrak{N},$$

from which we obtain, by scalar multiplication with suitable factors, that

$$a = \mathfrak{z}'\,\mathfrak{y}', \qquad b = \mathfrak{z}\mathfrak{z}' = \frac{1}{2}\,(\mathfrak{z}^2)' = 0, \qquad c = \mathfrak{z}'\,\mathfrak{N} = \frac{1}{\sqrt{g}}\,\mathfrak{z}'\,\mathfrak{y}'\,\mathfrak{z} = 0,$$

where the last equation follows from (3.34) and from the condition for developable surfaces. Therefore we have

$$\mathfrak{x}' = \mathfrak{y}'(1 + v(\mathfrak{z}'\,\mathfrak{y}')) + v'\,\mathfrak{z}.$$

The tangent of our curve is therefore in the direction of the generator $\mathfrak{z}$ provided that

$$1 + v(\mathfrak{z}'\,\mathfrak{y}') = 0. \tag{3.37}$$

If $\mathfrak{z}'\,\mathfrak{y}' = 0$ holds throughout a u^1-interval, it follows that $\mathfrak{z}' = 0$, so that all generators are parallel and the corresponding portion of the developable surface is a cylinder. Otherwise, we have an open portion of a surface in which $\mathfrak{z}'\,\mathfrak{y}' \neq 0$, and such open portions may be attached to each other along individual straight lines along which $\mathfrak{z}'\,\mathfrak{y}' = 0$. For the open portions, it follows from (3.37) that our curve in the surface satisfies

$$u^2 = v(u^1) = -\,1/\mathfrak{z}'\,\mathfrak{y}'.$$

Therefore there is at most one curve in the surface whose tangents have the directions of generators, and there is exactly one such curve for portions of the surface in which

$$0 \neq \mathfrak{x}' = \mathfrak{z}v'.$$

Finally, portions of the surface in which $v =$ const. satisfy $\mathfrak{x}(u^1) =$ const., so that all generators pass through a single point. Such portions of the surface are conical. We have, therefore, proved that *the most general developable surface is a ruled surface which consists of cylindrical portions, conical portions, and tangential developables, smoothly attached to each other along common generators.*

We proceed to the calculation of the metric coefficients of a developable surface. We have

$$\begin{aligned} \mathfrak{x}_1 &= \mathfrak{x}' + u^2\,\mathfrak{x}'' = \mathfrak{v}_1 + \kappa u^2\,\mathfrak{v}_2 \\ \mathfrak{x}_2 &= \mathfrak{x}' = \mathfrak{v}_1, \end{aligned} \tag{3.38}$$

so that

$$\begin{aligned} g_{11} &= \mathfrak{x}_1{}^2 = 1 + \kappa^2(u^1)\cdot(u^2)^2 \\ g_{12} &= g_{21} = \mathfrak{x}_1\,\mathfrak{x}_2 = 1 \\ g_{22} &= \mathfrak{x}_2{}^2 = 1. \end{aligned}$$

We see that these quantities are completely determined by the curvature κ of the edge of regression, whose torsion does not enter. For all space curves $\mathfrak{x}(u^1)$ (u^1, arc length) that have the same assigned curvature function $\kappa(u^1)$, the corresponding tangent surfaces have the same metric coefficients g_{ik} at a point with assigned parameter values u^i. If, on each of these surfaces, a curve is given by $u^i = u^i(t)$, it follows that the lengths

$$\int \sqrt{g_{ik}\,\dot{u}^i\,\dot{u}^k}\,dt$$

of corresponding pieces of these curves on the various surfaces agree. This remarkable state of affairs can also be expressed as follows: If we map one of the surfaces in question onto another by assigning to each of its points, with parameter values (u^i), the point on the other surface determined by the same parameter values, then this mapping preserves lengths, i.e., is *isometric*. Such an isometric mapping is also called an *application* and two surfaces that can be mapped isometrically onto each other are also called *applicable* (on each other). We can prove the following theorem:

Theorem 3.7.2. *The tangential developables are applicable on the plane.*

Proof. If $\kappa(u^1)$ is the curvature of the edge of regression of the given tangent surface, we can construct, according to the fundamental theorem for space curves, the plane curve having the same curvature function, that is, the plane curve determined by $\kappa(u^1)$ and $\tau = 0$ as its curvature and torsion, respectively. The tangent surface of this plane curve lies in the plane and is isometric to the given tangent surface because, as we have seen, its metric coefficients involve only the curvature, not the torsion.

The tangent surfaces are of special interest mainly because of their important property expressed by Theorem 3.7.2. We shall see later (II, §6.3) that conversely, the developable surfaces are the only surfaces applicable on the plane.

Exercise 3.7.3. Prove that all cones and cylinders are applicable on the plane. [*Hint.* A *cone* whose vertex is at the origin can be set up in the form

$$\mathfrak{x}(u^1, u^2) = u^2 \mathfrak{z}(u^1), \qquad \mathfrak{z}^2 = 1.$$

We can assume here that u^1 is the arc length along the curve $\mathfrak{z}(u^1)$, so that $\mathfrak{z}'^2 = 1$. Now begin by mapping the curve $u^2 = 1$ into the unit circle of the plane. For *cylinders*: $\mathfrak{x} = \mathfrak{y}(u^1) + u^2 \mathfrak{z}$, where $\mathfrak{z}$ is a constant unit vector.]

Exercise 3.7.4. *Prove* that a hyperboloid of one sheet contains two families of straight lines, and likewise for a hyperbolic paraboloid.

Exercise 3.7.5. *Prove* that a hyperboloid of revolution of one sheet is not applicable on the plane. (*Hint.* If the plane is mapped isometrically, its straight lines, being the shortest connections between any two of their points, must go into the straight lines on the hyperboloid, which are likewise shortest connections. Also, note that two generators of the hyperboloid that belong to the same family do not intersect.)

Exercise 3.7.6. *Prove*: No portion of a surface of a sphere is applicable on the plane. (*Hint.* We know from spherical trigonometry that its great circles are the shortest connections on the sphere. Thus, an isometric mapping would have to map them onto the straight lines of the plane. Furthermore, isometric mappings preserve angles because angles are measured in terms of the g_{ik}. Now, compare the sums of the angles in a triangle.)

Exercise 3.7.7. *Prove* that the ruled surface generated by (a) the principal normals, (b) the binormals of a space curve, is developable if and only if the curve is a plane curve. (*Hint.* Use the condition of Theorem 3.6.1 for developable surfaces, and Frenet's formulas.)

Exercise 3.7.8. Two surfaces $\mathfrak{x}(u^i)$, $\mathfrak{x}^*(u^i)$ that are applicable on each other are said to be transformable into each other by *bending* if there is a one-parameter family of isometric surfaces $\mathfrak{x}(u^i; t)$, $a \leq t \leq b$, which depends continuously on t and which is such that $\mathfrak{x}(u^i; a) = \mathfrak{x}(u^i)$ and $\mathfrak{x}(u^i; b) = \mathfrak{x}^*(u^i)$. *Show* that any developable surface can be bent into the plane. [*Hint* for the case of tangent surfaces: If κ, τ are curvature and torsion of the given edge of regression, then there is a family of edges of regression with curvatures $\kappa(u^1; t) = \kappa(u^1)$ and torsions $\tau(u^1; t) = t \cdot \tau(u^1)$.]

Exercise 3.7.9. *Prove* that the quadric surfaces mentioned in Exercise 3.7.4 are the only nonplanar surfaces that contain two families of straight lines. (This exercise does not presuppose any knowledge of differential geometry. *Hint*: Consider any three straight lines belonging to one of the families of generators. Then the straight lines of the second family are uniquely determined by the requirement of having to intersect these three lines. On the other hand, there is a quadric surface of one of the above-mentioned types that contains the three lines. Include in your discussion the cases in which the three lines are not mutually skew.)

Exercise 3.7.10. *Involutes of a Space Curve.* Generalizing the concept of involute of a plane curve (see Exercise 1.7.6), one calls an *involute* of a space curve $\mathfrak{x}(u^1)$ any curve in the tangent surface of $\mathfrak{x}(u^1)$ that intersects the tangents of $\mathfrak{x}(u^1)$ at right angles. Thus an involute satisfies

$$\mathfrak{x}^*(u^1) = \mathfrak{x}(u^1) + v(u^1)\mathfrak{x}'(u^1) \qquad \text{and} \qquad \mathfrak{x}'\,\mathfrak{x}^{*\prime} = 0.$$

Show that $|\mathfrak{x} - \mathfrak{x}^*| = \text{const.} \pm s$ and use this fact to justify the term "string involute" in a similar way as for plane curves.

Exercise 3.7.11. *Developable Surfaces as Envelopes of Families of Planes. Prove*: A one-parameter family of planes that is not part of a pencil of planes consists of the tangent planes of a developable surface. [*Hint.* Let t be the parameter of the family and let $\mathfrak{N}(t)$ be the unit normal vector of the plane, so that the planes of the family are given by the equation

$$\mathfrak{N}(t)\mathfrak{x} = c(t).$$

We may assume $\dot{\mathfrak{N}}(t) \neq 0$. The developable surface that is wanted then consists of all lines of intersection

$$\mathfrak{x}(u^1, u^2) = \mathfrak{x}(u^1) + u^2\,\mathfrak{x}'(u^1)$$

of the planes

$$\mathfrak{N}(t)\mathfrak{x} = c(t), \qquad \dot{\mathfrak{N}}(t)\mathfrak{x} = \dot{c}(t)$$

for all fixed $t = u^1$.]

Exercise 3.7.12. *Prove* that for every developable surface, $\det(L_{ik}) = 0$. [*Hint.* Use (3.29) and $\mathfrak{N}_2 = 0$.] Now use Exercise 3.5.1 to *determine* all surfaces of revolution that are developable.

Exercise 3.7.13. *Helicoid Ruled Surfaces.* If a straight line moves in such a way that each of its points moves on a circular helix, having a fixed straight line g as its axis, a ruled surface is generated which is called a helicoid ruled surface. Since every screw motion is composed of a rotation about the axis g through an angle φ and a translation by $h\varphi$ in the direction of g, such a motion can be represented in the form (making the x_3-axis coincide with g),

$$S(\mathfrak{x}) = \begin{pmatrix} \cos\varphi & -\sin\varphi & 0 \\ \sin\varphi & \cos\varphi & 0 \\ 0 & 0 & 1 \end{pmatrix} \mathfrak{x} + \begin{pmatrix} 0 \\ 0 \\ \varphi h \end{pmatrix}.$$

Setting $\varphi = u^2$ and taking $\mathfrak{x}_0 + u^1 \mathfrak{e}$ ($\mathfrak{e}^2 = 1$) as the initial position of the moving straight line, also choosing $\mathfrak{x}_0 = \{1, 0, 0\}$ which involves no loss of generality, we obtain the helicoid ruled surface in the form

$$\mathfrak{x}(u^1, u^2) = S(\mathfrak{x}_0 + u^1 \mathfrak{e}).$$

Investigate under what conditions this satisfies the requirements of the definition of surface ($\mathfrak{x}_1 \times \mathfrak{x}_2 \neq 0$) and show that the only helicoid ruled surfaces that are developable are the tangent surfaces of circular helices and the circular cylinders with g as axis. (Assume $h \neq 0$.) Investigate also the *winding surface* generated by a straight line that moves as above and intersects the axis g perpendicularly ($\mathfrak{e} = \{1, 0, 0\}$).

§ 4. Intrinsic Geometry of Surfaces

4.1 Introduction

We can describe our general plan for the further development of the theory of surfaces as follows. We shall make use of the decomposition of the curvature of surface curves into a tangential component κ_g and a normal component κ_n for the introduction of geometric quantities that are no longer dependent on the special choice of curves on the surface, but are determined solely by the geometry of the surface itself. In this section we take the first step in this direction, using arc length and the geodesic curvature κ_g.

The metric coefficients g_{ik} of a surface can be determined entirely by measurements within the surface and do not depend on any quantities outside the surface such as the normal vector. [In fact, g_{11} and g_{22} are obtained in terms of the arc lengths of the parametric curves, and g_{12}, in terms of the angle between them; see Eqs. (3.17) and (3.19).] Therefore, all quantities of the theory of surfaces that are expressible in terms of the g_{ik} alone can be determined by measurements within the surface. We saw that angles and areas (see §3.4) are examples of such quantities. In the case of the surface of the earth, for example, measurements "within the surface" would consist in measurements along the ground, while aerial pictures would be inadmissible. The part of surface theory that is concerned only with measurements within the surface is called the *intrinsic geometry* of surfaces and quantities determined entirely by the g_{ik} (and, thus, available through measurements within the surface) are called *intrinsic geometric* or *absolute* (or *geodesic*) quantities.

The intrinsic geometries of two mutually applicable surfaces are identical because the g_{ik} agree at corresponding points of the two surfaces. This implies that measurements within a surface cannot be sufficient to allow one to distinguish, for example, a plane from any other developable surface.

On the other hand, intrinsic geometric means are sufficient to allow one to distinguish between a portion of a spherical surface and a developable surface (see Exercise 3.7.6). Thus the intrinsic geometry of a surface does not nail down the trace of the surface in space, or, in other words, the quantities of intrinsic geometry do not constitute a complete system of invariants for surfaces. Nevertheless, intrinsic geometry deserves a particular place of interest in view of the further development of geometry that started with Riemann. The study of intrinsic geometry originated with Gauss (1827), whose motivation came partly from his position as surveyor. He observed, among other things, that many quantities that are at first defined in terms of measurements outside the surface (e.g., in terms of the normal vector) can be expressed in such a way that they are seen to be determined by the g_{ik} alone and are thus recognized as intrinsic quantities of the surface. In the following subsection we shall see that the geodesic curvature is such an intrinsic quantity.

4.2 The Geodesic Curvature as an Intrinsic Quantity

If

$$\mathfrak{x}(s) = \mathfrak{x}(u^i(s))$$

is a curve in a surface parametrized in terms of its arc length, then its geodesic curvature κ_g is given, according to (3.32), by

$$\kappa_g\, \mathfrak{s} = (u^{i''} + \Gamma^i_{jk}\, u^{j'}\, u^{k'})\mathfrak{x}_i, \qquad \mathfrak{s} = \mathfrak{N} \times \mathfrak{x}', \tag{4.1}$$

where the functions $\Gamma^i_{jk}(u^r)$ were defined in (3.27),

$$\mathfrak{x}_{jk} = \Gamma^i_{jk}\, \mathfrak{x}_i + L_{jk}\, \mathfrak{N}. \tag{4.2}$$

Forming the inner product of (4.2) with $\mathfrak{x}_l$ yields

$$\mathfrak{x}_{jk}\, \mathfrak{x}_l = \Gamma^i_{jk}\, \mathfrak{x}_i\, \mathfrak{x}_l = \Gamma^i_{jk}\, g_{il} \underset{\text{def}}{=} \Gamma_{jk|l}. \tag{4.3}$$

The quantities $\Gamma_{jk|l}$ and Γ^i_{jk} are called the *Christoffel symbols* of the first and second kind, respectively. They are expressible in terms of the functions g_{ik} alone, as follows. We have $g_{ik} = \mathfrak{x}_i\, \mathfrak{x}_k$. Differentiating with respect to u^l, we obtain

$$\frac{\partial g_{ik}}{\partial u^l} = \mathfrak{x}_{il}\, \mathfrak{x}_k + \mathfrak{x}_{kl}\, \mathfrak{x}_i = \Gamma_{il|k} + \Gamma_{kl|i}. \tag{4.4}$$

We note down also the corresponding equations obtained by permuting the indices,

$$\frac{\partial g_{il}}{\partial u^k} = \mathfrak{x}_{ik}\,\mathfrak{x}_l + \mathfrak{x}_{lk}\,\mathfrak{x}_i \tag{4.4'}$$

$$\frac{\partial g_{kl}}{\partial u^i} = \mathfrak{x}_{ki}\,\mathfrak{x}_l + \mathfrak{x}_{li}\,\mathfrak{x}_k, \tag{4.4''}$$

and we obtain from these, on account of the equality of certain pairs of second mixed partial derivatives, that

$$\frac{1}{2}\left(\frac{\partial g_{il}}{\partial u^k} - \frac{\partial g_{ik}}{\partial u^l} + \frac{\partial g_{lk}}{\partial u^i}\right) = \mathfrak{x}_{ik}\,\mathfrak{x}_l = \Gamma_{ik|l}. \tag{4.5}$$

Let us denote by (g^{ik}) the matrix inverse to (g_{ik}); it exists, since

$$\det(g_{ik}) = g > 0.$$

Thus we have

$$g^{il}\,g_{lk} = \delta^i_k, \qquad g^{il} = g^{li}; \tag{4.6}$$

we obtain

$$g^{11} = \frac{g_{22}}{g}, \qquad g^{12} = g^{21} = -\frac{g_{12}}{g}, \qquad g^{22} = \frac{g_{11}}{g}. \tag{4.7}$$

Multiplication of Eq. (4.3) by g^{lr} and summation over l yields

$$g^{lr}\,\Gamma_{jk|l} = \Gamma^i_{jk}\,g_{il}\,g^{lr} = \Gamma^i_{jk}\,\delta^r_i = \Gamma^r_{jk}. \tag{4.8}$$

Now, (4.8) and (4.5) together give

$$\Gamma^r_{ik} = \frac{g^{lr}}{2}\left(\frac{\partial g_{il}}{\partial u^k} - \frac{\partial g_{ik}}{\partial u^l} + \frac{\partial g_{lk}}{\partial u^i}\right). \tag{4.9}$$

As shown in (4.5), (4.8), and (4.7), the Christoffel symbols can be expressed in terms of the g_{ik} and their derivatives and, therefore, they are intrinsic quantities. We now return to the geodesic curvature. Inner multiplication of (4.1) by

$$\mathfrak{s} = \mathfrak{N} \times \mathfrak{x}' = \mathfrak{N} \times \mathfrak{x}_l\,u^{l'}$$

yields

$$\kappa_g = (u^{i''} + \Gamma^i_{jk}\,u^{j'}\,u^{k'})\,\mathfrak{N}\mathfrak{x}_l\,\mathfrak{x}_i\,u^{l'}. \tag{4.10}$$

We define

$$\varepsilon_{li} = \mathfrak{N}\mathfrak{x}_l\,\mathfrak{x}_i = \frac{(\mathfrak{x}_1 \times \mathfrak{x}_2)(\mathfrak{x}_l \times \mathfrak{x}_i)}{|\mathfrak{x}_1 \times \mathfrak{x}_2|}, \tag{4.11}$$

so that

$$\varepsilon_{11} = \varepsilon_{22} = 0, \qquad \varepsilon_{12} = -\,\varepsilon_{21} = |\mathfrak{x}_1 \times \mathfrak{x}_2| = \sqrt{g}. \tag{4.12}$$

Hence, (4.10) can be written

$$\kappa_g = \varepsilon_{li}\, u^{l'}(u^{i''} + \Gamma^i_{jk}\, u^{j'}\, u^{k'}). \tag{4.13}$$

Taking account of (4.9) and (4.12), we see that the right-hand side is expressible in terms of the g_{ik} and of the arc length of the curve itself, and we have proved the following:

Theorem 4.2.1. *The geodesic curvature is an intrinsic geometric quantity.*

The significance of formula (4.13) lies mainly in the fact that it contains the proof of this theorem. For many applications in the theory of surfaces, (4.13) proves to be inconvenient and, for this reason, we shall derive another, more convenient expression for κ_g, an expression that derives directly from the definition of κ_g as the tangential component of the curvature vector $\mathfrak{x}''$. Forming the inner product of (3.25) with $\mathfrak{s} = \mathfrak{N} \times \mathfrak{x}'$, we obtain

$$\kappa_g = \mathfrak{N}\mathfrak{x}'\,\mathfrak{x}''. \tag{4.14}$$

If we substitute in terms of the vectors of the moving trihedron of the curve and take account of $\mathfrak{v}_3 = \mathfrak{v}_1 \times \mathfrak{v}_2$, it follows that

$$\kappa_g = \mathfrak{N}(\mathfrak{x}' \times \mathfrak{x}'') = \kappa\mathfrak{N}\mathfrak{v}_3 \tag{4.15}$$

or, since the normal $\mathfrak{N}$ is perpendicular to the tangent plane of the surface, while the binormal $\mathfrak{v}_3$ of the curve is perpendicular to its osculating plane:

Theorem 4.2.2. *The geodesic curvature κ_g of a curve on the surface equals its curvature κ multiplied by the cosine of the angle between the osculating plane of the curve and the tangent plane of the surface.*

To be sure, this property does not exhibit the intrinsic geometric nature of geodesic curvature, since it makes essential use of properties of the curve in space.

Exercise 4.2.3. *Prove*: If a surface is a plane, then the geodesic curvature κ_g of any of its curves is the same as the ordinary curvature κ.

Exercise 4.2.4. *Determine* the geodesic curvature of a circle of latitude on a sphere. (*Hint.* Use Theorem 4.2.2.)

Exercise 4.2.5. *Prove* that a curve of constant geodesic curvature on a sphere is a circle.

Exercise 4.2.6. *Show* that the ε_{ik} of Eq. (4.11) are transformed, under a parameter transformation with positive functional determinant, in the same way as are the g_{ik}.

Exercise 4.2.7. *Geodesic Curvature of Parametric Curves.* For the curves u^1 = const., i.e., for the u^2-curves, the relations $u^{1\prime} = u^{1\prime\prime} = 0$ (with s_2 as arc length) imply that the geodesic curvature κ_{g^2} as given by (4.13) is

$$\kappa_{g^2} = \varepsilon_{21}\, u^{2\prime}\, \Gamma^1_{22}(u^{2\prime})^2 = -\sqrt{g}\,\Gamma^1_{22}\left(\frac{du^2}{ds_2}\right)^3.$$

Since $ds_2{}^2 = g_{22}(du^2)^2$, this becomes

$$\kappa_{g^2} = -\frac{\sqrt{g}}{\sqrt{g_{22}}^{\,3}}\,\Gamma^1_{22}$$

and correspondingly we obtain for the geodesic curvature κ_{g^1} of the u^1-curves that

$$\kappa_{g^1} = (\sqrt{g}/\sqrt{g_{11}}^{\,3})\,\Gamma^2_{11}.$$

In particular, if the parametric net is orthogonal, so that $g_{12} = 0$ and therefore $g = g_{11}\, g_{22}$, then the Christoffel symbols of Eq. (4.9) become

$$\Gamma^1_{22} = -\frac{1}{2g_{11}}\frac{\partial g_{22}}{\partial u^1}, \qquad \Gamma^2_{11} = -\frac{1}{2g_{22}}\frac{\partial g_{11}}{\partial u^2}$$

and finally,

$$\kappa_{g^2} = \frac{d}{ds_1}\log\sqrt{g_{22}}, \qquad \kappa_{g^1} = -\frac{d}{ds_2}\log\sqrt{g_{11}},$$

where s_1, s_2 denote arc length along the u^1-curves and along their orthogonal trajectories, the u^2-curves, respectively.

Exercise 4.2.8. Let $\Theta(s)$ be the angle that a curve $\mathfrak{x}(s)$ on the surface makes with the u^1-curve of an orthogonal parametric net. *Prove* the following result, due to Liouville (1850):

$$\kappa_g = \frac{d\Theta}{ds} + \kappa_{g^1}\cos\Theta + \kappa_{g^2}\sin\Theta.$$

(*Hint.* Use Exercise 4.2.7.)

4.3 The Geodesic Curves (Curves with Geodesic Curvature Zero)

In the plane, the curves of vanishing geodesic curvature — i.e., the straight lines — play a distinguished role. It is to be expected that, in the case of an arbitrary surface, the curves on the surface whose geodesic curvature vanishes ($\kappa_g = 0$), the so-called *geodesic curves* or *geodesics*, play a similarly important part, especially since, for curves on the surface, Exercise 4.2.3 shows the geodesic curvature to be a generalization of the curvature of plane curves.

For geodesics, we deduce from (4.1) and (4.14) the following result:

Theorem 4.3.1. *The geodesics on a surface are characterized by each of the following two conditions:*

$$u^{i''} + \Gamma^i_{jk} u^{j'} u^{k'} = 0 \tag{4.16}$$

and

$$\mathfrak{N}\mathfrak{x}' \mathfrak{x}'' = 0, \tag{4.17}$$

where the parameter is arc length.

Condition (4.16) shows that the property of a curve of being a geodesic is an intrinsic geometric property. This implies that, under an application of a surface onto another surface, geodesics are mapped onto geodesics. Condition (4.17) shows that if $\mathfrak{x}'' \neq 0$, then the geodesics are precisely those curves whose osculating plane contains the normal to the surface.

Theorem 4.3.2. *For every point $\mathfrak{x}(\underset{0}{u}^i)$ of a surface and for every direction on the surface through that point, there is one and only one geodesic through the given point in the given direction.*

Proof. Let the direction on the surface be given by its unit vector $\mathfrak{v} = v^i \mathfrak{x}_i$, with $\mathfrak{v}^2 = g_{ik}(\underset{0}{u}^i)v^i v^k = 1$. Then the geodesic must be a solution of the system (4.16) of differential equations,

$$u^{i''} = - \Gamma^i_{jk}(u^r) u^{j'} u^{k'}$$

with the initial conditions $u^i(0) = \underset{0}{u}^i$, $u^{i'}(0) = v^i$.

From the existence and uniqueness theorem for systems of ordinary differential equations, it follows that there exists one and only one solution $u^i(s)$ satisfying the initial conditions. This settles the uniqueness assertion of Theorem 4.3.2. We still have to show that the unique solution actually gives a geodesic. To this end, it suffices to show that s is indeed the arc length of the curve defined by $u^i(s)$, since the assertion then follows from (4.16). Toward this, in turn, we consider the function $f(s) = g_{ik}(u^i(s))u^{i'}u^{k'}$ and we will show that $f'(s) = 0$. As the initial condition $g_{ik}(\underset{0}{u}^j)v^i v^k = f(0) = 1$ then yields $f(s) = 1$, we will have shown that s is the arc length. Now

$$f'(s) = \frac{\partial g_{ik}}{\partial u^j} u^{j'} u^{i'} u^{k'} + g_{ik} u^{i''} u^{k'} + g_{ik} u^{i'} u^{k''}.$$

In accordance with (4.4) and (4.3), we set

$$\frac{\partial g_{ik}}{\partial u^j} = \Gamma_{ij|k} + \Gamma_{kj|i} = g_{kr} \Gamma^r_{ij} + g_{ir} \Gamma^r_{kj}, \tag{4.18}$$

from which it follows after suitably renaming the indices of summation that

$$f'(s) = g_{kr}(u^{r''} + \Gamma^r_{ij} u'^i u'^j) u^{k'} + g_{ir}(u^{r''} + \Gamma^r_{kj} u^{k'} u^{j'}) u^{i'}.$$

The expressions in parentheses vanish since the functions $u^i(s)$ are solutions of the system (4.16); therefore, $f'(s) = 0$.

4.4 The Extremal Property of the Geodesic Curves

A straight line of the plane is characterized by its property of being the shortest possible connecting curve between any two of its points. We shall show that the geodesic curves on a surface are characterized by a corresponding property.

Let $u^i(t)$, $t_0 \leqq t \leqq t_1$, be a curve on a surface which is the shortest possible curve on this surface connecting its two end points $\mathfrak{x}(t_0)$, $\mathfrak{x}(t_1)$. Let $u^i(t; \varepsilon)$ be a family of curves on the surface which, for $|\varepsilon| < \tau$ and $t_0 \leqq t \leqq t_1$, satisfy $u^i(t; 0) = u^i(t)$ and $u^i(t_0; \varepsilon) = u^i(t_0)$ and $u^i(t_1; \varepsilon) = u^i(t_1)$. All of these curves connect the two given end points. Therefore, if the shortest connection is to be the member of the family for which $\varepsilon = 0$, then the lengths $L(\varepsilon)$ of these curves must satisfy the condition $L'(0) = 0$.

For the length $L(\varepsilon)$, we obtain, setting $u^i = u^i(t; \varepsilon)$:

$$L(\varepsilon) = \int_{t_0}^{t_1} \sqrt{g_{ik}(u^j) \dot{u}^i \dot{u}^k}\, dt \underset{\text{def}}{=} \int_{t_0}^{t_1} w\, dt. \tag{4.19}$$

Differentiating the integrand w with respect to ε, we obtain

$$\frac{\partial w}{\partial \varepsilon} = \frac{1}{2w} \left(\frac{\partial g_{ik}}{\partial u^j} \frac{\partial u^j}{\partial \varepsilon} \dot{u}^i \dot{u}^k + 2 g_{ij} \dot{u}^i \frac{\partial^2 u^j}{\partial t\, \partial \varepsilon} \right).$$

We take $u^i(t; \varepsilon) = u^i(t) + \varepsilon v^i(t)$, where $v^i(t)$ satisfies $v^i(t_0) = v^i(t_1) = 0$, and assume (without loss of generality) that t is the arc length of the curve $u^i(t; 0)$; then the radical w has the value 1 at $\varepsilon = 0$ and we have $\partial u^i / \partial \varepsilon = v^i$ (see Fig. 15). It follows that

$$L'(0) = \frac{1}{2}\int_{t_0}^{t_1}\left(\frac{\partial g_{ik}}{\partial u^j}\, v^j\, \dot{u}^i\, \dot{u}^k + 2g_{ij}\, \dot{u}^i\, \dot{v}^j\right) dt.$$

We integrate the second term of the integrand by parts and we obtain, observing that $v^i(t)$ vanishes at the ends of the interval of integration,

$$L'(0) = \frac{1}{2}\int_{t_0}^{t_1}\left(\frac{\partial g_{ik}}{\partial u^j}\, \dot{u}^i\, \dot{u}^k - 2\,\frac{\partial(g_{ij}\, \dot{u}^i)}{\partial t}\right) v^j\, dt.$$

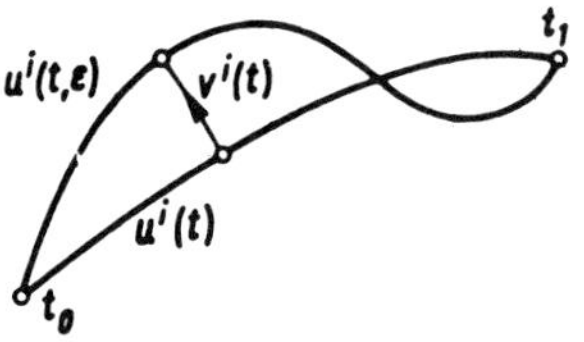

Fig. 15. Geodesic curve.

Since $L'(0) = 0$ must hold for all admissible $v^i(t)$, it follows from the fundamental lemma of the calculus of variations (see Appendix II.3) that

$$\frac{1}{2}\,\frac{\partial g_{ik}}{\partial u^j}\, \dot{u}^i\, \dot{u}^k - \frac{\partial(g_{ij}\, \dot{u}^i)}{\partial t} = 0$$

or

$$\frac{1}{2}\,\frac{\partial g_{ik}}{\partial u^j}\, \dot{u}^i\, \dot{u}^k - \frac{\partial g_{ij}}{\partial u^k}\, \dot{u}^k\, \dot{u}^i - g_{rj}\, \ddot{u}^r = 0.$$

We break up the second term and obtain, after renaming the indices of summation,

$$g_{rj}\, \ddot{u}^r + \Gamma_{ik|j}\, \dot{u}^i\, \dot{u}^k = 0$$

which, after multiplication with g^{lj} and summation over j, yields

$$\ddot{u}^l + \Gamma^l_{ik}\, \dot{u}^i\, \dot{u}^k = 0.$$

Since the parameter t is arc length, we have arrived at the system (4.16) which characterizes the geodesic curves, so that we have proved the following theorem:

Theorem 4.4.1. *If a curve on a surface is the shortest connection on the surface between its two end points, then this curve is (a part of) a geodesic curve.*

The condition of being a geodesic is, of course, not sufficient for a curve to be a shortest connection, since the derivative of $L(\varepsilon)$ vanishes also in the case of a maximum, or, more generally, in the case of any stationary value. And, in fact, simple examples show that not every arc of a geodesic curve on a surface is the shortest connection between its end points. A directrix (circle) on a circular cylinder or a great circle on a sphere are geodesic curves on account of (4.17), but arcs of these curves that exceed a semicircle are certainly not the shortest connection between their end points. However, we shall prove in 4.6 that sufficiently short geodesic arcs do have the minimal property of being shortest connections.

Exercise 4.4.2. *Prove* that the meridians of a surface of revolution are geodesics. Hence the totality of geodesics of the sphere consists of the great circles. [*Hint.* Use (4.17).]

Exercise 4.4.3. *Prove* that if a surface contains a straight line, that straight line is a geodesic curve on the surface.

Exercise 4.4.4. *Prove*: If it is possible to chose an orthogonal parametric net on a surface in such a way that the parametric curves are geodesics, then the surface is applicable on the plane. (*Hint.* Use Exercise 4.2.7.)

Exercise 4.4.5. *Show* that the torsion of a geodesic equals $\mathfrak{x}' \mathfrak{N}\mathfrak{N}'$. For any curve on a surface, the term *geodesic torsion* denotes the torsion of the geodesic curve that is tangent, at the point in question, to the given curve. Give examples showing that this quantity is not intrinsic. (*Hint.* Use Exercise 4.4.6.)

Exercise 4.4.6. *Prove* that the geodesic curves on a circular cylinder are circular helices, including the limiting cases of the generators and of the circles perpendicular to them. (*Hint.* If the cylinder is applied on the plane, its geodesics must be mapped onto straight lines.)

Exercise 4.4.7. *Derive* the following characterization of geodesics on a surface of revolution: If β is the angle between the tangent to the curve and a circle of latitude and if r is the radius of this circle, then $r \cdot \cos \beta = \text{const.}$ along the curve (A. C. Clairaut, 1735). (*Hint.* Using the notation of Exercise 3.3.1, we have $\cos \beta = \mathfrak{x}_2 \mathfrak{x}' / |\mathfrak{x}_2| = \mathfrak{x}_2 \mathfrak{x}' / r$.) The result can be used to give another proof for Exercise 4.4.6.

Exercise 4.4.8 *Prove*: If a geodesic is parametrized by a parameter t, not necessarily the arc length, then

$$\ddot{u}^i + \Gamma^i_{jk} \dot{u}^j \dot{u}^k = \lambda \cdot \dot{u}^i \qquad \text{with} \qquad \lambda = -\frac{d^2 t}{ds^2} \cdot \left(\frac{ds}{dt}\right)^2.$$

Prove also: A necessary and sufficient condition for a curve $u^i(t)$ to be a geodesic curve is that

$$\ddot{u}^1 \dot{u}^2 - \ddot{u}^2 \dot{u}^1 + (\dot{u}^2 \Gamma^1_{ik} - \dot{u}^1 \Gamma^2_{ik}) \dot{u}^i \dot{u}^k = 0.$$

4.5 Geodesics Nets

A special type of parameters, associated with so-called *geodesic nets*, is often useful.

Definition. *A surface is said to be parametrized in terms of geodesic parameters $\bar{u}^i$, also called geodesic parallel coordinates, if the $\bar{u}^1$-curves are geodesics and the $\bar{u}^2$-curves are orthogonal trajectories of these geodesics.*

Since a geodesic net is orthogonal, we have $\bar{g}_{12} = 0$. The geodesic curvature κ_{g^1} of the $\bar{u}^1$-curves vanishes, which together with Exercise 4.2.7 implies that $\partial \bar{g}_{11}/\partial \bar{u}^2 = 0$. Hence $\bar{g}_{11}$ is a function of $\bar{u}^1$ alone: $\bar{g}_{11} = g_{11}(\bar{u}^1)$.

Now we introduce new parameters by a parameter transformation whose allowability we leave for the reader to verify, as an exercise:

$$u^1 = \int^{\bar{u}^1} \sqrt{g_{11}(\bar{u}^1)}\, d\bar{u}^1, \qquad u^2 = \bar{u}^2.$$

The new parameter u^1 is the arc length of the u^1-lines. The element of arc length is given, in terms of the new parameters, by

$$ds^2 = (du^1)^2 + g_{22}(u^i)\,(du^2)^2. \tag{4.20}$$

Thus, by merely transforming the parameter of the $\bar{u}^1$-curves, we can always arrive at geodesic parameters for which $g_{11} = 1$, $g_{12} = 0$.

FIG. 16. Geodesic parallel coordinates.

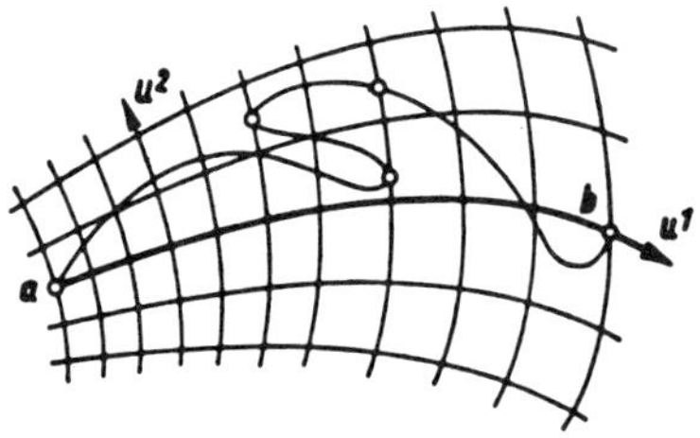

FIG. 17. Geodesics as shortest curves.

Now let C be a curve on the surface. We construct a geodesic net whose u^1-curves are the geodesics orthogonal to C and whose u^2-curves are the orthogonal trajectories of these geodesics (see Fig. 16). Then C is itself a parametric curve, say the curve $u^1 = 0$. The arc length of a portion of a geodesic between C and a curve $u^1 = c_0$ is given, by (4.20), by

$$\int_0^{c_0} du^1 = c_0,$$

which is independent of u^2. This establishes the following result:

Theorem 4.5.1. *The curves* $u^1 =$ const. *of a geodesic net are geodesic parallels, meaning that, for two fixed curves of this family, their distance from each other as measured along the geodesic u^1-curves remains constant.*

In general, if we start with a given curve C and construct the corresponding geodesic net as above, this geodesic net will give an admissible parametrization of the surface only in a strip around C which is covered simply (i.e., without overlappings) and completely by the geodesics that intersect C orthogonally.

4.6 The Minimal Property of the Geodesic Curves

Using geodesic nets, we can now prove the minimal property for sufficiently short arcs of geodesic curves. We recall that a family of curves is called a *field* if its members give a simple covering of a region G, called the domain of the field. We have the following theorem, due to Gauss:

Theorem 4.6.1. *If a geodesic arc belongs to a field of geodesics, then it is the shortest connection of its end points that is possible within the domain G of the field.*

Proof. We choose the geodesics of the field and their orthogonal trajectories to be the parametric curves of a geodesic net, so that the element of arc is given by (4.20). Let $u^2 = u^2(u^1)$, $a \leqq u^1 \leqq b$, give a piecewise smooth curve in G which connects the end points of the given geodesic arc, with $u^2(a) = u^2(b) = 0$. By (4.20), the length of this curve satisfies

$$\left| \int_a^b \sqrt{1 + g_{22}\, u_2'^2}\, du^1 \right| \geqq \left| \int_a^b du^1 \right| = |b - a|,$$

which proves the theorem in the restricted sense that only a restricted class of curves has so far been admitted for comparison with the given geodesic arc. To complete the proof, we must also admit curves that may not be representable by an equation $u^2 = u^2(u^1)$ and which may, therefore, intersect a curve $u^1 =$ const. more than once. We can partition such a curve into subarcs, each of which is representable by a single-valued function $u^2(u^1)$ (see Fig. 17), and we can then finish the argument as above.

One cannot dispense with the imbedding in a field as is shown by the examples discussed at the end of §4.4.

Exercise 4.6.2. *Geodesic Polar Coordinates.* If, for the u^1-curves, we choose the geodesics through a fixed point P and, for the u^2-curves, their orthogonal trajectories, we obtain a special type of geodesic net, the geodesic polar coordinates with center (pole) P. By Theorem 4.5.1, the u^2-curves can also be described as those curves along which the distance from P is fixed when measured along the u^2-curves; in particular, therefore, they are closed curves. *Prove* that the parameter u^2 can be chosen as the angle which the geodesics through P make with a fixed direction on the surface at P. (*Hint.* The point P itself is a singular point for this parametric net. In the proof, start with curves at distance $\varepsilon > 0$ from the point P and then let ε tend to 0.)

4.7 Geodesic Parallelism

We have seen, in the last subsection, how to find "geodesic parallels" of a given curve on a surface. However, this concept of parallelism for curves on the surface does not have all the properties that attach to the corresponding concept in the plane. The parallel curves of a straight line in the plane are themselves straight lines, but the geodesic parallels of a geodesic on a surface are themselves geodesics if and only if the surface is applicable on the plane. (This follows from Exercise 4.4.4.) This fact at first raises the suspicion that it is not possible to give a reasonable extension of the concept of parallelism from the plane to arbitrary surfaces. However, in 1917 — thus, very late in the history of the theory of surfaces — Levi-Cività discovered an important concept that permits a definition of parallelism for vectors in a surface.

Let $\mathfrak{x}(t) = \mathfrak{x}(u^i(t))$ be a curve on a surface and let $\mathfrak{v}(t) = \mathfrak{x}_i\, v^i(t)$ be a vector field defined along this curve and consisting of vectors tangent to the surface. For the special case of the plane, the vectors of this field are parallel if the condition $\dot{\mathfrak{v}} = 0$ holds. Following Levi-Cività, we set down the following definition in the case of an arbitrary surface:

Definition. *Let $\mathfrak{x}(u^i(t))$ be a curve on a surface and let $\mathfrak{v}(t)$ be a vector field defined on this curve. Then the vectors of the vector field are called geodesically parallel along (or with respect to) the curve, if the tangential component of $\dot{\mathfrak{v}}(t)$ vanishes identically in t.*

For such a vector field, it follows from Eq. (4.3) that

$$0 = \mathfrak{x}_i\, \dot{\mathfrak{v}} = \mathfrak{x}_i \frac{d}{dt}(\mathfrak{x}_k\, v^k) = \mathfrak{x}_i\, \mathfrak{x}_k\, \dot{v}^k + \mathfrak{x}_i\, \mathfrak{x}_{kj}\, \dot{u}^j\, v^k = g_{ik}\, \dot{v}^k + \Gamma_{kj|i}\, \dot{u}^j\, v^k,$$

whence, by multiplication by g^{ir} and summation over i, we obtain, as an equivalent condition, the two equations

$$\dot{v}^r + \Gamma^r_{kj}\, \dot{u}^j\, v^k = 0. \qquad (4.21)$$

Formula (4.21) contains, beyond the vector field $v^i(t)$ and the curve $u^i(t)$, only the quantities Γ^i_{jk}, which are expressible in terms of the g_{ik}. Therefore, the concept of geodesic parallelism of vectors with respect to a curve belongs to the intrinsic geometry of the surface, which justifies the adjective "geodesic." Once again, we are dealing with a concept for whose definition we made essential use of the imbedding of the surface in space, but which then turned out to depend only on measurements within the surface. The fact that the concept is also independent of any particular parametrization of the surface follows from its definition, since the requirement that the vector $\dot{\mathfrak{v}}$ should be perpendicular to the surface is invariant under parameter transformations.

We note that (4.21) is a system of two ordinary linear differential equations for the two functions $v^i(t)$. The existence and uniqueness theorem for such systems therefore implies the following result:

Theorem 4.7.1. *Let $u^i(t)$ be a curve on a surface and let $\underset{0}{v}^i$ be a vector tangent to the surface at the point $u^i(t_0)$. Then there exists one and only one vector field $v^i(t)$ that consists of geodesically parallel tangent vectors with respect to the curve and that satisfies $v^i(t_0) = \underset{0}{v}^i$.*

Let us call a curve $u^i(t)$ on the surface *maximally straight* if the tangent vectors of the curve constitute a field of geodesically parallel vectors with respect to the curve. From (4.21) it follows that, for such a curve, we have

$$\ddot{u}^r + \Gamma^r_{kj}\,\dot{u}^j\,\dot{u}^k = 0,$$

which proves the following result:

Theorem 4.7.2. *The geodesic curves on a surface are also the maximally straight curves on the surface.*

Thus we have found yet another property that the geodesic curves on a surface share with the straight lines of the plane.

An ordinary translation (parallel displacement) in the plane preserves lengths and angles. We now state and prove an analogous result for geodesic parallelism.

Theorem 4.7.3. *Let $u^i(t)$ be a curve on a surface and let each of $v^i(t)$, $w^i(t)$ be a field of geodesically parallel vectors with respect to $u^i(t)$. Then the angle between v^i and w^i is constant along the curve, and the vectors $v^i(t)$ are all of equal length.*

Proof. We have

$$\frac{d}{dt}(g_{ik}(u^j(t))v^i w^k) = \frac{\partial g_{ik}}{\partial u^j}\dot{u}^j v^i w^k + g_{ik}\dot{v}^i w^k + g_{ik} v^i \dot{w}_k,$$

and, since the parallelism condition (4.21) for the two vector fields gives

$$\dot{v}^i = -\Gamma^i_{mr} v^m \dot{u}^r, \qquad \dot{w}^k = -\Gamma^k_{lr} w^l \dot{u}^r,$$

it follows upon suitable re-naming of summation indices that

$$\frac{d}{dt}(g_{ik} v^i w^k) = \left(\frac{\partial g_{ml}}{\partial u^r} - \Gamma_{mr|l} - \Gamma_{lr|m}\right)\dot{u}^r v^m w^l = 0$$

by (4.4). This proves first that lengths remain constant (apply the last equation to the special case $v^i = w^i$) and then that angles remain constant, as was to be proved.

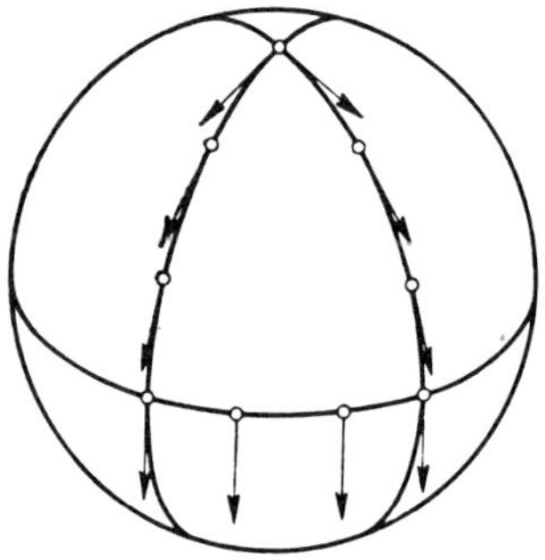

FIG. 18. Parallel vector field on the sphere.

We re-emphasize that geodesic parallelism of vectors is only defined *with respect to* (*or along*) *a curve.* Indeed, two vectors attached to two points of a surface may be geodesically parallel as members of a field along one curve joining the two points on the surface, *without* being geodesically parallel with respect to a second such curve. See, in this connection, Exercise 4.7.4 below.

Exercise 4.7.4. *Geodesically Parallel Vector Fields along Curves on a Sphere* (see Fig. 18). Let N be the north pole of a sphere and let P_1, P_2 be two points on its equator. *Prove*: The vector field that consists of all unit vectors pointing south at all points of the three geodesic arcs NP_1, P_1P_2, P_2N is a field of geodesic parallels. In particular, therefore, every unit vector at the north pole can be moved into any other unit vector at the north pole by means of geodesic parallel displacement along suitable curves. (*Hint.* Use Theorems 4.7.2 and 4.7.3.)

Exercise 4.7.5. *Let* $v^i(s)$ be a field of geodesically parallel vectors along a curve $u^i(s)$ on a surface. Let κ_g be the geodesic curvature of this curve and let β be the angle between the tangent vector $\dot{u}^i$ and the field vector v^i. *Prove* that $\kappa_g = d\beta/ds$. (*Hint.* The curve parameter s may be taken to be arc length, without loss of generality.)

Solution. By Theorem 4.7.3, two fields of parallel vectors form a constant angle, so that there is no loss of generality in assuming $v^i(s_0) = u^{i\prime}(s_0)$ at the point in question. Setting $\mathfrak{v}(s) = v^i\,\mathfrak{x}_i$, we then have

$$\sin \Delta\beta = (\mathfrak{v}(s_0 + \Delta s) \times \mathfrak{x}'(s_0 + \Delta s)) \cdot \mathfrak{N}(s + \Delta s)$$

$$= (\mathfrak{x}'(s_0) + \Delta s\mathfrak{v}'(s_0)) \times (\mathfrak{x}'(s_0) + \Delta s\mathfrak{x}''(s_0)) \cdot [\mathfrak{N}(s_0) + \Delta s\mathfrak{N}'(s_0)] + \Delta s^2(\ldots).$$

Now

$$\mathfrak{v}' = (v^i\,\mathfrak{x}_i)' = -\,\Gamma^i_{jk}\,v^j\,u^{k\prime}\,\mathfrak{x}_i + v^i\,\mathfrak{x}_{ik}\,u^{k\prime},$$

and, since $v^i = u^{i\prime}$ holds at the point in question, it follows from (4.2) that

$$\mathfrak{v}' = L_{ik}\,u^{i\prime}\,u^{k\prime}\,\mathfrak{N}.$$

Substituting this in the expression for $\sin \Delta\beta$ above, we obtain

$$\frac{d\beta}{ds} = \lim_{\Delta s \to 0} \frac{\Delta\beta}{\Delta s} = \lim_{\Delta s \to 0} \frac{\sin \Delta\beta}{\Delta s} = \mathfrak{x}'\,\mathfrak{x}''\,\mathfrak{N} = \kappa_g.$$

Exercise 4.7.6. *The Invariant Differential of a Vector Field.* Let $\mathfrak{v} = \mathfrak{v}(u^i)$ be a differentiable vector field defined on a surface. The tangential component of $d\mathfrak{v}$ is called the *invariant differential* of the field and is written $D\mathfrak{v}$. In particular, let $\mathfrak{v}$ be a field of tangent vectors to the surface: $\mathfrak{v} = v^i(u^j)\mathfrak{x}_i$, so that $d\mathfrak{v} = dv^i\,\mathfrak{x}_i + v^i\,\mathfrak{x}_{ik}\,du^k$. If the quantities Dv^i are defined by

$$D\mathfrak{v} = \mathfrak{x}_i\,Dv^i,$$

then, considering that $D\mathfrak{v}$ is the tangential component of $d\mathfrak{v}$, we have $\mathfrak{x}_k\,d\mathfrak{v} = \mathfrak{x}_k\,D\mathfrak{v} = \mathfrak{x}_k\,\mathfrak{x}_i\,Dv^i = g_{ik}\,Dv^i$, hence

$$Dv^j = g^{jk}\,\mathfrak{x}_k\,d\mathfrak{v} = dv^j + \Gamma^j_{rk}\,v^r\,du^k.$$

By its definition, the invariant differential is invariant under parameter transformations, and the last expression given for it shows it to be an intrinsic geometric quantity of the surface. In terms of it, one can define the *covariant derivative* $v^i{}_{;k}$ of the vector field v^i, by means of $Dv^i = v^i{}_{;k}\,du^k$. Thus we have

$$v^i{}_{;k} = \frac{\partial v^i}{\partial u^k} + \Gamma^i_{rk}\,v^r.$$

This covariant derivative plays a central role in the intrinsic geometry of surfaces. We shall study it when we get to Riemannian geometry.

Exercise 4.7.7. If two surfaces are tangent along a curve, then this curve has the same geodesic curvature on both surfaces (for, $\mathfrak{x}'$ and $\mathfrak{x}''$ are determined by the curve itself, while $\mathfrak{N}$ is the same for both surfaces). Hence, if a curve on a surface is given, then a developable surface that is tangent to the surface along this curve can be circumscribed about the surface. (The developable surface is the envelope of the tangent planes to the surface along the curve.) This developable surface can be applied on the plane. The geodesic curvature of the given curve is then equal to the curvature of the corresponding plane curve. Use this fact for an alternative proof in Exercise 4.7.5.

§ 5. Curvature of Surfaces

This section takes up the geometric interpretation of the normal component of the curvature of curves on a surface, its tangential component having been studied in § 4 above.

5.1 The Second Fundamental Form and the Two Principal Curvatures

Let $\mathfrak{x}(t) = \mathfrak{x}(u^i(t))$ be a curve on a surface and let s be its arc length. As in the definition (3.25), we have

$$\mathfrak{x}'' = \kappa \mathfrak{v}_2 = \kappa_n \mathfrak{N} + \kappa_g \mathfrak{N} \times \mathfrak{x}', \tag{5.1}$$

where the prime (′) denotes, as before, differentiation with respect to the arc length s while the dot (·) will denote differentiation with respect to the parameter t. Using the coefficients

$$L_{ik} = \mathfrak{N}\mathfrak{x}_{ik} = -\mathfrak{N}_i \mathfrak{x}_k$$

of the second fundamental form, we obtain from (5.1), (3.24), and (3.31) that

$$\kappa_n = \mathfrak{x}'' \mathfrak{N} = \kappa \mathfrak{v}_2 \mathfrak{N} = L_{ik} u^{i\prime} u^{k\prime} = \frac{L_{ik} \dot{u}^i \dot{u}^k}{g_{ik} \dot{u}^i \dot{u}^k}. \tag{5.2}$$

This contains the following result:

Theorem 5.1.1. *All curves on a surface that pass through a given point and share the same tangent line at this point, also must share the same normal curvature κ_n at the point.*

Therefore to determine this normal curvature, it is sufficient to obtain the normal curvature of the curve of intersection of the surface with the plane perpendicular to the tangent plane and through the given tangential direction at the point. This plane normal section has geodesic curvature zero at the point in question, since it contains the corresponding vector $\mathfrak{x}''$, which is perpendicular to $\mathfrak{x}'$ and which, therefore, is in the direction of the surface normal $\mathfrak{N}$. Hence, for this plane normal section, the absolute value of its normal curvature coincides with that of its ordinary curvature κ, or, in other words,

Theorem 5.1.2. *Let κ_n be the normal curvature of a curve on a surface, at a point P_0, and let κ be the curvature of the plane normal section at P_0 that is tangent at P_0 to the given curve. Then $|\kappa_n| = \kappa$.*

Theorem 5.1.1 shows that, unlike geodesic curvature, the normal curvature depends only on the direction through the given point on the surface. We may therefore expect that the totality of the normal curvatures at a point of the surface bears on the curvature behavior of the surface itself at the point.

The part $\kappa_n = \kappa \mathfrak{N}\mathfrak{v}_2$ of Eq. (5.2) can also be expressed as follows:

Theorem 5.1.3 (Meusnier, 1776). *The normal curvature of a curve on a surface equals the product of its curvature by the cosine of the angle α between the principal normal of the curve and the normal of the surface.*

Exercise 5.1.4. *Prove* the following version, important for descriptive geometry, of Meusnier's theorem 5.1.3: For a family of surface curves that are mutually tangent at a point P of the surface, the centers of curvature all lie on a circle lying in the common normal plane of these curves, this circle being tangent to the surface at P and of diameter $1/|\kappa_n|$. (*Hint.* See Fig. 19.)

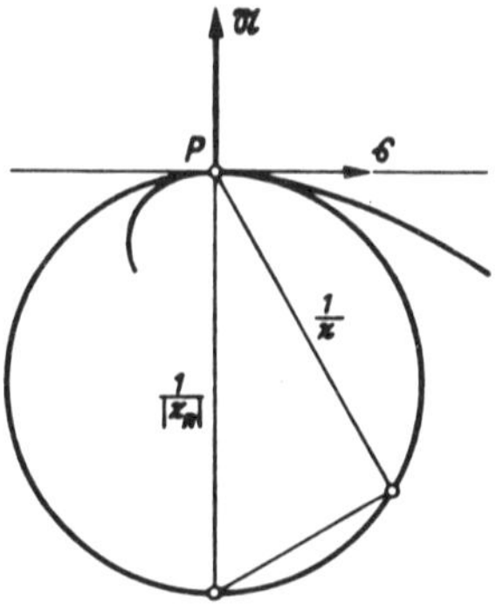

FIG. 19. Illustrating Meusnier's theorem.

Next, let P_0 be a fixed point on the surface, and let g_{ik} and L_{ik} be the coefficients of the first and of the second fundamental form, respectively. Let us determine, among the directions through the point P_0, those in which κ_n assumes values that are extrema. By Theorem 5.1.1, we can, without loss of generality, represent all directions through P_0 by means of unit vectors $u^{i\prime} = \xi^i$, so that $g_{ik}\, \xi^i\, \xi^k = 1$. Then we are to determine the extrema of $L_{ik}\, \xi^i\, \xi^k$ [see (5.2)] subject to the constraints $g_{ik}\, \xi^i\, \xi^k = 1$. The method of Lagrange's multipliers leads to the conditions

$$\frac{\partial}{\partial \xi^i} (L_{jk}\, \xi^j\, \xi^k - \lambda g_{jk}\, \xi^j\, \xi^k) = 0 \tag{5.3}$$

or

$$(L_{ik} - \lambda g_{ik})\xi^k = 0. \tag{5.4}$$

The system (5.4) of homogeneous linear equations has nontrivial solutions if and only if

$$0 = \det(L_{ik} - \lambda g_{ik}) = L - \lambda(g_{11} L_{22} - 2g_{12} L_{12} + g_{22} L_{11}) + \lambda^2 g, \tag{5.5}$$

where

$$L \underset{\text{def}}{=} \det(L_{ik}), \qquad g \underset{\text{def}}{=} \det(g_{ik}). \tag{5.6}$$

The quadratic equation (5.5) in λ has two solutions κ_1, κ_2. Both of these are real; either this can be proved by some linear algebra (see Exercise 5.1.8, below), or it can be deduced from the fact that κ_n, being a continuous function on a closed bounded set (the circle $|\xi^i \mathfrak{x}_i| = 1$), must assume its maximum and its minimum. Thus κ_1, κ_2 furnish the extrema of κ_n.

We denote corresponding solutions of (5.4), satisfying the constraints, by $\xi^i_{(1)}, \xi^i_{(2)}$. Hence by (5.2), (5.4),

$$\kappa_\mu = L_{ik}\, \xi^i{}_{(\mu)}\, \xi^k{}_{(\mu)} \qquad \text{with} \qquad g_{ik}\, \xi^i_{(\mu)}\, \xi^k_{(\mu)} = 1$$

for $\mu = 1, 2$, so that the κ_μ are precisely the normal curvatures in the directions $\xi^i_{(\mu)}$ through P_0. The two numbers κ_1, κ_2 are called the *principal curvatures* of the surface at the point P_0. By (5.5), they satisfy the following relations, applying Viëta's rules for the roots:

$$H \underset{\text{def}}{=} \frac{\kappa_1 + \kappa_2}{2} = \frac{g_{11} L_{22} - 2g_{12} L_{12} + g_{22} L_{11}}{2g} = \frac{1}{2} g^{ik} L_{ik} \tag{5.7}$$

and

$$K \underset{\text{def}}{=} \kappa_1 \kappa_2 = \frac{L}{g}. \tag{5.8}$$

The number H, being the arithmetic mean of the principal curvatures, is called the *mean curvature* of the surface at P_0; the number K, their product, is called the *Gaussian curvature* at P_0.

For the corresponding *directions of principal curvature*, $\xi^i_{(\mu)} \mathfrak{x}_i$ (also called *principal directions* or *curvature directions* at P_0), we have by (5.4) that

$$(L_{ik} - \kappa_1 g_{ik})\xi^k_{(1)} = 0, \qquad g_{ik}\, \xi^i_{(1)}\, \xi^k_{(1)} = 1 \tag{5.9$_1$}$$

and

$$(L_{ik} - \kappa_2 g_{ik})\xi^k_{(2)} = 0, \qquad g_{ik}\, \xi^i_{(2)}\, \xi^k_{(2)} = 1. \tag{5.9$_2$}$$

If we multiply (5.9_2) by $\xi^i_{(1)}$, sum over i, and substitute into (5.9_1), we obtain

$$0 = (\kappa_1 - \kappa_2) g_{ik}\, \xi^i_{(1)}\, \xi^k_{(2)}, \tag{5.10}$$

which shows that if $\kappa_1 \neq \kappa_2$ then the two principal directions are mutually perpendicular. If $\kappa_1 = \kappa_2$, then (5.5) has a double real root and it follows that $L_{ik} = \kappa_1 g_{ik}$ (see Exercise 5.1.8, below), whence it follows that Eq. (5.4) is satisfied for all directions ξ^i on the surface through P_0; in this case, all the normal curvatures at P_0 are equal, and such points are called *umbilical points* (or *umbilics*, or *navel points*) of the surface. Every direction through an umbilic could be called a principal direction. In particular, we can arbitrarily select any pair of mutually perpendicular directions through an umbilic and label them the principal directions. We can summarize what we have just found, as follows:

Theorem 5.1.5. *At each point of a surface, there are two mutually perpendicular principal directions in the surface such that the normal curvature assumes its maximum in the one direction and its minimum in the other.*

Now assume that the parametrization of the surface has been chosen in such a way that the two parametric curves through P_0 are tangent at P_0 to the two principal directions. Then by Theorem 5.1.5, we have $g_{12} = 0$ at P_0. Further, since $\xi^2_{(1)} = \xi^1_{(2)} = 0$ at P_0, it follows from (5.9) that

$$L_{11} = \kappa_1 g_{11}, \qquad L_{22} = \kappa_2 g_{22}, \qquad L_{12} = 0.$$

We can assume without loss of generality that $g_{11} = g_{22} = 1$ at P_0. Then, if a direction in the surface at P_0 is specified by ξ^i satisfying

$$g_{ik}\, \xi^i \xi^k = (\xi^1)^2 + (\xi^2)^2 = 1,$$

we can set $\xi^1 = \cos\varphi$, $\xi^2 = \sin\varphi$, where φ is the angle which the specified direction makes with the first principal direction. Substitution of the above in (5.2) leads to the following result:

Theorem 5.1.6. (Euler, 1760). *Let κ_n be the normal curvature of a curve on a surface and let φ be the angle between the curve and the first principal direction. Then*

$$\kappa_n = \kappa_1 \cos^2\varphi + \kappa^2 \sin^2\varphi. \tag{5.11}$$

The name *line of curvature* is given to a curve on a surface if, at every point of the curve, its normal curvature is equal to one of the principal curvatures or, in other words, if the curve is everywhere tangent to one of the principal directions. Through every point other than an umbilic, there are exactly two lines of curvature; they correspond to solutions of the following differential equation, obtained from (5.4) by eliminating λ:

$$\begin{vmatrix} L_{1k}\,du^k & g_{1k}\,du^k \\ L_{2k}\,du^k & g_{2k}\,du^k \end{vmatrix} = 0. \tag{5.12}$$

It seems plausible on geometric grounds that on portions of a surface free of umbilics or flat points (these being points where all of the L_{ik} vanish), the lines of curvature can be chosen to serve as the parametric curves. We shall prove that this is indeed the case.

Theorem 5.1.7. *Let P_0 be a point on a surface that is neither an umbilic nor a flat point. Then there is a neighborhood of P_0 on the surface and a parametrization $\bar{u}^i$ of that neighborhood such that the parametric curves are the lines of curvature.*

Proof. First, by a standard continuity argument, there is a neighborhood of P_0 on the surface that is free of umbilics. The differential equation (5.12) can be written

$$B_{ik}\,du^i\,du^k = 0,$$

where

$$B_{11} = L_{11}\,g_{12} - g_{11}\,L_{12}, \qquad 2B_{12} = L_{11}\,g_{22} - g_{11}\,L_{22}, \qquad B_{22} = L_{12}\,g_{22} - g_{12}\,L_{22}.$$

Using the notation of Theorem B, §3.3, we have

$$\det(B_{ik}) = -\,g^2(H^2 - K),$$

as can be verified by working out each side. The right-hand side is negative at points other than umbilics. Now, Theorem B of §3.3 establishes the desired conclusion. Furthermore, if $\mathfrak{x}(u^i)$ is of class C^k then the g_{ik} are of class C^{k-1} and the L_{ik} of class C^{k-2}; hence the B_{ik} are also of class C^{k-2}. Therefore, by the theorem just referred to, the functions $\bar{u}^i(u^j)$, which give the transformation to the lines-of-curvature parametrization, are likewise of class C^{k-2}.

Exercise 5.1.8. *Prove* that the two roots of (5.5) are always real and that they coincide if and only if $L_{ik} = c g_{ik}$. (*Hint.* By means of an affine transformation, the matrices L_{ik} and g_{ik} can be transformed into diagonal form such that $g_{ik} = \delta_{ik}$; this last requirement can be met, since the quadratic form $g_{ik}\,\xi^i\,\xi^k$ is positive-definite. This affine transformation corresponds to a parameter transformation for the surface, and in the new parameters we have $L_{11} = L_1$, $L_{12} = 0$, $L_{22} = L_2$, $g_{ik} = \delta_{ik}$.)

Exercise 5.1.9. *Prove*: We always have $H^2 \geqq K$, with equality at umbilics and nowhere else.

Exercise 5.1.10. *Prove*: On the sphere, all points are umbilics.

Exercise 5.1.11. *Prove*: The mean curvature H at any point is the average value of the normal curvatures at this point; that is, $H = (1/2\pi)\int_0^{2\pi} \kappa_n \, d\varphi$ (where φ has the same meaning as in Theorem 5.1.6, which may also be used in the proof).

Exercise 5.1.12. *Prove*: For any two mutually perpendicular directions at a point, the arithmetic mean of the normal curvatures in these directions equals the mean curvature H at the point.

Exercise 5.1.13. *Prove*: A necessary and sufficient condition for the parametric curves to be lines of curvature is that $g_{12} = L_{12} = 0$. [*Hint.* The necessity follows from the proof of Theorem 5.1.5. To prove the sufficiency of the condition, use (5.5) and (5.2).]

Exercise 5.1.14. *Prove* that the lines of curvature of a surface of revolution are its meridians and circles of latitude. (*Hint.* Use Exercise 5.1.13.)

Exercise 5.1.15. Give examples (for instance, circular cylinders) showing that κ_n and H are not invariant under bending, also examples showing that the notion of umbilical point is not an intrinsic-geometric notion.

5.2 The Dupin Indicatrix

We shall describe a device that is helpful for gaining some geometric insight into curvature relations near a point $\mathfrak{x}_0 = \mathfrak{x}(\underset{0}{u}^i)$, $\underset{0}{u}^i = 0$, on a surface. The device consists in carrying out a small translation of the tangent plane

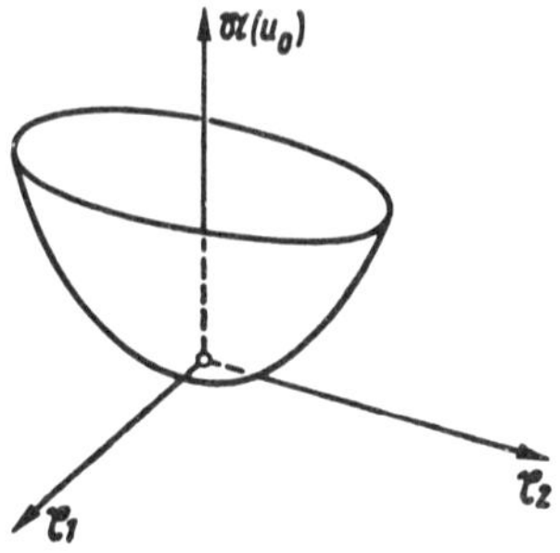

FIG. 20. Deriving the indicatrix.

at $\mathfrak{x}_0$ and then investigating its curve of intersection with the surface. Let $\mathfrak{x}_0 + \varepsilon\mathfrak{N}(0)$ be the point of intersection of the normal to the surface at $\mathfrak{x}_0$ with the translated plane (parallel to the tangent plane). Then the equation of this translated plane (see Fig. 20) is

$$\varepsilon = \mathfrak{N}(\underset{0}{u}^i)\,(\mathfrak{x} - \mathfrak{x}_0). \qquad (5.13)$$

Hence, its intersection with the surface is given by

$$\varepsilon = \mathfrak{N}(\underset{0}{u^i})\,(\mathfrak{x}(u^i) - \mathfrak{x}_0). \tag{5.14}$$

Now, from Taylor's formula, we have

$$\mathfrak{x}(u^i) - \mathfrak{x}_0 = \mathfrak{x}_i(0)u^i + \tfrac{1}{2}\,\mathfrak{x}_{ik}(0)u^i u^k + R_3. \tag{5.15}$$

Substituting (5.15) into (5.14) and noting that $\mathfrak{N}\mathfrak{x}_i(0) = 0$, $\mathfrak{N}\mathfrak{x}_{ik} = L_{ik}$, we obtain

$$\varepsilon = \tfrac{1}{2}\,L_{ik}(0)u^i u^k + R_3, \tag{5.16}$$

and, writing $\xi^i = u^i/\sqrt{2|\varepsilon|}$, this becomes

$$\pm 1 = L_{ik}(0)\xi^i \xi^k + R_3/\varepsilon. \tag{5.17}$$

Now we let ε tend to 0 and observe that $\lim_{\varepsilon\to 0} R_3/\varepsilon = 0$, so that we obtain the following locus in the tangent plane:

$$L_{ik}(0)\xi^i \xi^k = \pm 1. \tag{5.18}$$

Definition. *The locus* (5.18), *which is obtained by magnification of the curve of intersection in the ratio* $1:\sqrt{2\,|\varepsilon|}$ *and passage to the limit as* $\varepsilon \to 0$, *is called the Dupin indicatrix at* $\mathfrak{x}_0$.

Theorem 5.2.1. *The Dupin indicatrix is*

(1) *an ellipse with center* 0 *if the second fundamental form* $L_{ik}\,\xi^i\,\xi^k$ *is definite,*

(2) *a pair of conjugate hyperbolas if the form is indefinite,*

(3) *a pair of straight lines if the form is degenerate,*

(4) *not defined if all* L_{ik} *vanish.*

The point $\mathfrak{x}_0$ *of the surface is called in these four cases an elliptic point, hyperbolic point, parabolic point, or flat point, respectively.*

The principal directions (of curvature) coincide with the directions of the axes of the Dupin indicatrix.

For the lengths of the half-axes, we obtain from Euler's theorem 5.1.6, choosing coordinates suitably as in that theorem,

$$\frac{1}{\sqrt{|\kappa_1|}} \quad \text{and} \quad \frac{1}{\sqrt{|\kappa_2|}} \tag{5.19}$$

The Dupin indicatrix provides information about the behavior of the surface in the neighborhood of the point in question (see Fig. 21), as follows:

1. *Elliptic Points.* A sufficiently small neighborhood of the point P_0 in question lies entirely on one side of the tangent plane at P_0, since, for sufficiently small u^i, equation (5.16) implies that ε has the (fixed) sign of the definite second fundamental form. Since $L = \det(L_{ik}) > 0$, it follows that $K = L/g > 0$. As examples of surfaces all of whose points are elliptic, we mention ellipsoids and hyperboloids of two sheets.

The indicatrix is a circle if and only if P_0 is an umbilic.

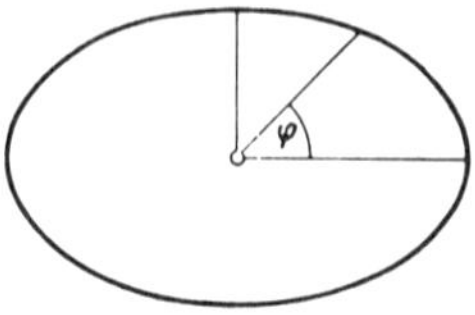

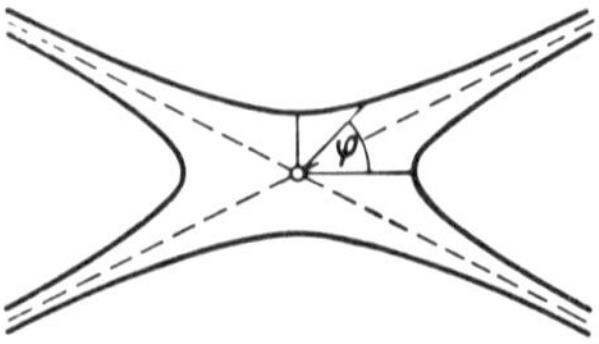

FIG. 21. The Dupin indicatrix at

(a) an elliptic point, (b) a hyperbolic point,

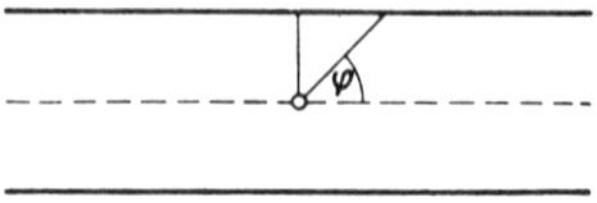

(c) a parabolic point, and the root of the radius of normal curvature $1/\sqrt{|\kappa_n|}$, in an arbitrary direction φ.

2. *Hyperbolic Points.* The tangent plane at P_0 intersects the surface in two curves whose tangents at P_0 are the asymptotes of the indicatrix. For, to obtain these curves of intersection $u^i(t)$, $u^i(0) = 0$, we set $\varepsilon = 0$ in (5.16), and it follows from

$$0 = \frac{1}{2} L_{ik}(0) \frac{u^i}{t} \frac{u^k}{t} + \frac{R_3}{t^2}$$

that $L_{ik} \dot{u}^i \dot{u}^k = 0$, since R_3 goes to zero of order three in t. The integral curves of this differential equation are tangent at every one of their points to an asymptote of the indicatrix at the point, and are called the *asymptotic curves* of the surface; their directions are the *asymptotic directions* at any one of their points. The angle between them is bisected by the principal directions, since the latter are along the axes of the indicatrix.

At hyperbolic points we have $L < 0$, hence $K < 0$. As examples of surfaces all of whose points are hyperbolic, we mention the hyperboloids of one sheet.

3. *Parabolic Points and Flat Points.* At such points the rank of the matrix (L_{ik}) is less than 2. If this rank equals 1 then the Dupin indicatrix is a pair of parallel straight lines $[L_{11}(\xi^1)^2 = 1$ in a suitably chosen coordinate system$]$. If it equals 0, then the indicatrix cannot even be defined; this is the case of a flat point, so called because the tangent plane at such a point has contact of at least second order with (or "osculates") the surface. At parabolic points, including flat points, we have $K = 0$.

The developable surfaces are examples of surfaces that consist entirely of parabolic points (possibly including flat points). The plane consists only of flat points.

Exercise 5.2.2. *Prove* that the asymptotic curves on a hyperboloid of one sheet are its straight lines.

Exercise 5.2.3. C. Dupin introduced his indicatrix, in 1813, in a somewhat different way. He considered an osculating paraboloid, meaning a paraboloid that has contact of at least second order with the surface at P_0. In our notation (with $u^1 = x_1$, $u^2 = x_2$), such a paraboloid is given by

$$\mathfrak{y}(u^i) = \mathfrak{x}_0 + \mathfrak{x}_i(0)u^i + \tfrac{1}{2}\,\mathfrak{x}_{ik}(0)u^i\,u^k,$$

since the difference $\mathfrak{y}(u^i) - \mathfrak{x}(u^i)$ vanishes at $u^i = 0$, as do its first and second derivatives. The plane sections of the paraboloid, parallel to the tangent plane, are similar and homothetic to the indicatrix. The paraboloid is elliptic or hyperbolic or a parabolic cylinder, according to whether the point P_0 is an elliptic or hyperbolic or parabolic point.

Exercise 5.2.4. Two directions on a surface at P_0 are called *conjugate directions at* P_0 if they belong to conjugate diameters of the indicatrix. *Prove* that a necessary and sufficient condition for the parametric curves to form a conjugate net (i.e., to have two conjugate tangents at every point) is that $L_{12} = 0$.

Exercise 5.2.5. *Prove*: (a) The parametric curves of a surface are its asymptotic curves if and only if $L_{11} = L_{22} = 0$. (b) On a sufficiently small portion of a surface consisting of hyperbolic points only, the asymptotic curves can be chosen to serve as the parametric curves. (*Hint.* The differential equation of the asymptotic curves is $L_{ik}\,du^i\,du^k = 0$. Use Theorem B of § 3.3 and include a discussion of the differentiability conditions, as was done in the proof of Theorem 5.1.7.)

Exercise 5.2.6. *Prove*: A curve on a surface that is not a straight line is an asymptotic curve of the surface if and only if its osculating plane coincides with the tangent plane of the surface. (*Hint.* Use Meusnier's theorem and note that the osculating plane is spanned by the principal normal and the tangent line of the curve.)

5.3 Geometric Interpretation of the Gaussian Curvature

The quantities H and K were introduced in connection with curvature measurements at a point of a surface, but in a rather formal way — as combinations of the principal curvatures. It would be desirable to introduce some more direct geometric measurement of curvature on the surface, which we proceed to do, following Gauss.

If we attempt to extend to the case of surfaces any of the geometric definitions of the curvature of a curve, we find that one of these permits an immediate extension, namely the one that defines curvature in terms of the ratio of the length of the spherical tangent image to that of the curve itself (see Exercise 1.7.4; for plane curves, alternatively, the ratio of the length of the spherical principal normal image to that of the curve itself, see Exercise 2.3.6). Following Gauss, let us map the surface onto the unit sphere by means of "parallel normals": $\mathfrak{x}(u^i) \to \mathfrak{N}(u^i)$. Let P_0 be a point

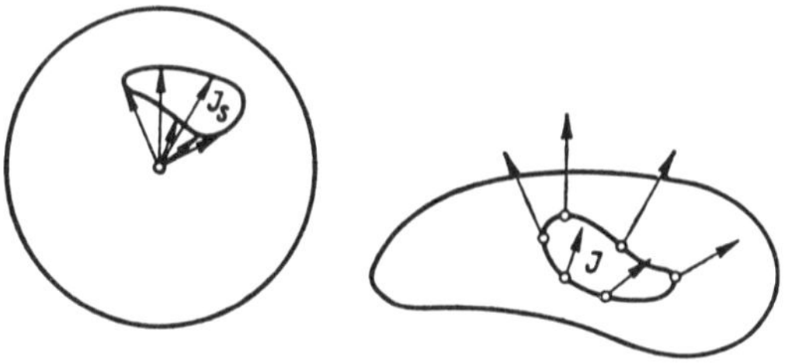

FIG. 22. Spherical normal image.

of the given surface and let $\mathfrak{U}$ be a parameter region such that the corresponding points of the surface constitute a neighborhood of P_0 on the surface. The mapping by parallel normals (see Fig. 22) assigns to this neighborhood a portion of the unit sphere which we can parametrize with the same parameters (and thus refer to the same parameter region $\mathfrak{U}$). Let J be the area of the neighborhood of P_0 and let J_s be the (signed) area of the spherical normal image of J. Their ratio is

$$\frac{J_s}{J} = \frac{\iint\limits_{\mathfrak{U}} \mathfrak{N}\mathfrak{N}_1\mathfrak{N}_2 \, du^1 \, du^2}{\iint\limits_{\mathfrak{U}} \mathfrak{N}\mathfrak{x}_1\mathfrak{x}_2 \, du^1 \, du^2}. \tag{5.20}$$

This quotient will be the larger the more the surface is "curved" near P_0, since the normals will then spread out over a larger part of the unit sphere. Letting the neighborhood of P_0 shrink down to P_0, we obtain

$$\lim \frac{J_s}{J} = \frac{\mathfrak{N}\mathfrak{N}_1 \mathfrak{N}_2}{\mathfrak{N}\mathfrak{x}_1 \mathfrak{x}_2}. \tag{5.21}$$

This number depends only on the point P_0 of the surface; Gauss introduced it as a measure of curvature. Toward obtaining a simpler expression for it, we shall calculate the derivatives of the normal vector in terms of the vectors $\mathfrak{N}$, $\mathfrak{x}_1$, $\mathfrak{x}_2$ of the moving trihedron of the surface. We set

$$\mathfrak{N}_k = B^i_k\,\mathfrak{x}_i + C_k \cdot \mathfrak{N}. \tag{5.22}$$

The coefficients B^i_k, C_k are determined as follows: First, differentiation of $\mathfrak{N}^2 = 1$ gives $\mathfrak{N} \cdot \mathfrak{N}_k = 0$, so that scalar multiplication of (5.22) by $\mathfrak{N}$ yields $C_k = 0$. Next, scalar multiplication of (5.22) by $\mathfrak{x}_j$ yields, by (3.29),

$$L_{jk} = -\,B^i{}_k\,g_{ij}.$$

Multiplying this by g^{jr} and summing over j, we have

$$B^r{}_k = -\,g^{rj}\,L_{jk}.$$

Thus, we have expressed the coefficients in question in terms of those of the two fundamental forms. Substituting into (5.22), we obtain *Weingarten's equations*

$$\mathfrak{N}_k = -\,L^i{}_k\,\mathfrak{x}_i \qquad \text{with} \qquad L^i{}_k \underset{\text{def}}{=} g^{ij}\,L_{jk}. \tag{5.23}$$

Substitution from these into the numerator of (5.21) yields

$$\lim J_s/J = \frac{\mathfrak{N}L^k{}_1\,\mathfrak{x}_k\,L^j{}_2\,\mathfrak{x}_j}{\mathfrak{N}\mathfrak{x}_1\,\mathfrak{x}_2} = \frac{(\mathfrak{N}\mathfrak{x}_1\,\mathfrak{x}_2)(L^1_1\,L^2_2 - L^1_2\,L^2_1)}{\mathfrak{N}\mathfrak{x}_1\,\mathfrak{x}_2} = \det(L^i{}_k).$$

Now (5.23) and the multiplication theorem for determinants show that

$$\det(L^i{}_k) = \det(g^{ik})\,\det(L_{ik}) = \frac{L}{g} = K, \tag{5.24}$$

so that we have proved

Theorem 5.3.1. *The Gaussian curvature K at P_0 equals*

$$\lim \frac{J_s}{J},$$

that is, the limit of the ratio between the area of the spherical (normal) image of a neighborhood of P_0 and the area of this neighborhood itself, where the limit is taken as this neighborhood shrinks down to P_0.

Note that this reveals a new geometric significance of the sign of K, which we already know to be decisive for the elliptic or hyperbolic character of the point: K is positive or negative according as the spherical mapping preserves or reverses orientation.

Exercise 5.3.2 *Prove*: If the parametric curves are the lines of curvature, and in that case only, Weingarten's equations (5.23) become the *formulas of Rodrigues*,

$$\mathfrak{N}_i = -\kappa_i \mathfrak{x}_i \qquad (\textit{not} \text{ to be summed over } i!).$$

(*Hint.* Use Exercise 5.1.13.)

Exercise 5.3.3. *Prove*: A curve on a surface is a line of curvature if and only if the surface normals at its points generate a developable surface. (*Hint.* Use either Theorem 3.6.1 and Eq. (5.12) or Exercise 5.3.2.)

Exercise 5.3.4. *Prove* that the sign of H has no geometric meaning; it depends on the sign of $\mathfrak{N}$.

Exercise 5.3.5. *Prove* the following theorem (Beltrami and Enneper, 1870): The torsion of an asymptotic curve, other than a straight line, is $\pm\sqrt{-K}$. [*Hint.* Along an asymptotic curve, the unit binormal vector $\mathfrak{v}_3$ coincides with the unit normal vector of the surface (see Exercise 5.2.6); hence, differentiation with respect to the arc length s yields, using the third Frenet formula,

$$\mathfrak{N}' = \mathfrak{v}_3{}' = -\tau\mathfrak{v}_2$$

or $\tau = -\mathfrak{N}'\,\mathfrak{v}_2 = -\mathfrak{N}'(\mathfrak{v}_3 \times \mathfrak{v}_1) = -\mathfrak{N}'(\mathfrak{N} \times \mathfrak{x}')$. Therefore, Weingarten's equations yield

$$\tau = L_k^i\, \mathfrak{x}_i\, u^{k\prime}(\mathfrak{N} \times \mathfrak{x}') = L_k^i\, u^{k\prime}\, u^{j\prime}(\mathfrak{x}_i\, \mathfrak{N}\mathfrak{x}_j) = (L_k^2\, u^{k\prime}\, u^{1\prime} - L_k^1\, u^{k\prime}\, u^{2\prime})\sqrt{g}.$$

Here, substitute from the following relations, which are consequences of the differential equation of the asymptotic curves: $L_{1k}\, u^{k\prime} = \pm\, u^{2\prime}\sqrt{-L}$ and $L_{2k}\, u^{k\prime} = \mp\, u^{1\prime}\sqrt{-L}$ (where the upper signs hold for the one asymptotic curve, the lower signs for the other). Then use $L_k^j = g^{jr}\, L_{rk}$ and $g^{11} = g_{22}/g$, etc., and the fact that s is the arc length, so that $g_{ik}\, u^{i\prime}\, u^{k\prime} = 1$.]

5.4 Formulas of Gauss and Weingarten

In the theory of space curves, we developed formulas (the Frenet formulas) for the derivatives of the vectors of the moving trihedron, and we arrived at a theorem to the effect that the coefficients in these formulas — namely, curvature and torsion — constitute a complete system of invariants for a curve. The corresponding questions in the case of surfaces suggest that we write down the derivatives of the vectors $\mathfrak{N}$, $\mathfrak{x}_1$, $\mathfrak{x}_2$ of the moving trihedron in terms of these three vectors. The derivatives are given by the *formulas of Gauss*, (3.27), and those of *Weingarten*, (5.23), which we recapitulate here:

$$\begin{aligned} \mathfrak{x}_{ik} &= \Gamma_{ik}^r\, \mathfrak{x}_r + L_{ik}\, \mathfrak{N} \\ \mathfrak{N}_i &= -L^r{}_i\, \mathfrak{x}_r \end{aligned} \tag{5.25}$$

with

$$\Gamma_{ik}^{r} = \frac{1}{2} g^{rl} \left(\frac{\partial g_{li}}{\partial u^k} - \frac{\partial g_{ik}}{\partial u^l} + \frac{\partial g_{lk}}{\partial u^i} \right), \qquad L^r{}_i = g^{rj} L_{ji}. \tag{5.26}$$

Equations (5.25) correspond to the Frenet formulas of the theory of space curves. We shall see in §11.7 that the coefficients in (5.25) constitute a complete system of invariants for the surface. Here we shall deduce from these equations, which are basic throughout the entire theory of surfaces, some other important consequences.

The third derivatives $\mathfrak{x}_{ikj}$ satisfy the relations

$$\mathfrak{x}_{ikj} = \mathfrak{x}_{ijk}; \tag{5.27}$$

these we shall now make use of. Denoting differentiation with respect to u^j by $\ldots,_j$, we obtain from (5.25) by differentiation that

$$\mathfrak{x}_{ikj} = \Gamma_{ik,j}^{r} \mathfrak{x}_r + \Gamma_{ik}^{r} \mathfrak{x}_{rj} + L_{ik,j} \mathfrak{N} + L_{ik} \mathfrak{N}_j.$$

Here, we substitute for the derivatives of $\mathfrak{x}_r$ and $\mathfrak{N}$ from (5.25), after suitable renaming of subscripts, and we thus obtain

$$\mathfrak{x}_{ikj} = (\Gamma_{ik,j}^{l} + \Gamma_{ik}^{r} \Gamma_{rj}^{l} - L_{ik} L_j^l) \mathfrak{x}_l + (L_{ik,j} + \Gamma_{ik}^{r} L_{rj}) \mathfrak{N}. \tag{5.28}$$

We obtain a corresponding expression for $\mathfrak{x}_{ijk}$ by interchanging k and j, so that (5.27) and (5.28) then yield

$$0 = G^l{}_{ikj} \mathfrak{x}_l + C_{ikj} \mathfrak{N} \tag{5.29}$$

with

$$\begin{aligned} G^l{}_{ikj} &= R^l{}_{ikj} - (L_{ik} L_j^l - L_{ij} L_k^l) \\ R^l{}_{ikj} &= \Gamma_{ik,j}^{l} - \Gamma_{ij,k}^{l} + \Gamma_{ik}^{r} \Gamma_{rj}^{l} - \Gamma_{ij}^{r} \Gamma_{rk}^{l} \\ C_{ikj} &= L_{ik,j} - L_{ij,k} + \Gamma_{ik}^{r} L_{rj} - \Gamma_{ij}^{r} L_{rk}. \end{aligned} \tag{5.30}$$

Since the triple $\mathfrak{N}$, $\mathfrak{x}_1$, $\mathfrak{x}_2$ is linearly independent, every one of the coefficients in (5.29) must vanish. Thus we have the following relations, due to Gauss:

$$G^l{}_{ikj} = 0, \tag{5.31}$$

and the following further relations, due to Mainardi and Codazzi,[1]

[1] Discovered by Mainardi in 1856, and independently by Codazzi in 1860. The relations of Gauss are to be found, like most of his contributions to the theory of surfaces, in his *Disquisitiones generales circa superficies curvas* of 1827.

$$C_{ikj} = 0. \tag{5.32}$$

Theorem 5.4.1. *The two fundamental forms* g_{ik} *and* L_{ik} *of a surface are related by Eqs.* (5.31) *and* (5.32).

5.5 The Theorema egregium

Let us make use of Gauss' equations (5.31). By (5.30), they amount to the same as

$$R^r{}_{ikj} = L_{ik} L^r_j - L_{ij} L^r_k = g^{rl}(L_{ik} L_{jl} - L_{ij} L_{kl})$$

which, multiplied by g_{mr} and summed over r, yields

$$g_{mr} R^r{}_{ikj} = L_{ik} L_{jm} - L_{ij} L_{km}.$$

For $i = k = 1$, $j = m = 2$, the right-hand side is $L = \det(L_{ik})$, while the left-hand side involves only quantities which, by (5.30), can be expressed in terms of the g_{ik} (and their derivatives up to the second order) and which are therefore intrinsic-geometric quantities. Since $K = L/g$ holds for the Gaussian curvature K, we have thus proved the following theorem which Gauss himself called extraordinary (*egregius* meaning extraordinary or excellent):

Theorem 5.5.1. (*Theorema egregium* of Gauss.) *The Gaussian curvature* K *is a quantity that belongs to the intrinsic geometry of the surface and which is therefore, in particular, invariant under bending.* K *can be expressed in terms of the* g_{ik} *and their derivatives, as follows:*

$$K = \frac{g_{2j}}{g} R^j{}_{112}.$$

Gauss called attention to this property of K, by his epithet, because it does express a remarkable fact; K was introduced in various ways — as the product of the principal curvatures, as the quotient L/g, as the limit of a ratio of areas in connection with the spherical mapping by parallel normals — in all of which essential use was made of the imbedding of the surface in space, in particular, of the normal vector. Yet this quantity then turned out to be obtainable by means of measurements within the surface; that is, it turned out to be intrinsic. Thus the situation is similar to what we found earlier in connection with the geodesic curvature, or with geodesic parallelism.

Exercise 5.5.2. Use the *theorema egregium* for a new proof of the fact that no sphere, or portion thereof, is applicable on the plane.

Exercise 5.5.3. *Prove*: The character of a point (elliptic, or hyperbolic, or parabolic including flat points) is invariant under bending. In particular, developable surfaces consist entirely of parabolic points and flat points.

Exercise 5.5.4. *Prove* that if the parametric net is orthogonal ($g_{12} = 0$), then

$$K = -\frac{1}{2\sqrt{g}}\left\{\left(\frac{g_{11,2}}{\sqrt{g}}\right)_{,2} + \left(\frac{g_{22,1}}{\sqrt{g}}\right)_{,1}\right\}.$$

Exercise 5.5.5. *Prove*: If the parametric net is geodesic ($g_{12} = 0$, $g_{11} = 1$), then

$$\frac{\partial^2 \sqrt{g_{22}}}{(\partial u^1)^2} + K\sqrt{g_{22}} = 0.$$

Exercise 5.5.6. *Prove*: If the parametric curves are the lines of curvature, then Eqs. (5.32) of Mainardi and Codazzi become

$$L_{ii\ k} = \frac{1}{2}\, g_{ii,k}\left(\frac{L_{11}}{g_{11}} + \frac{L_{22}}{g_{22}}\right) = g_{ii,k} \cdot H \qquad (i \neq k, = 1, 2).$$

§ 6. Special Topics in the Theory of Surfaces

6.1 Preliminary Remarks on Method

In §3–§5 we have developed the local theory of surfaces in Euclidean space in its essential features. To bring this theory to a certain state of completion still requires the following up of two important lines of ideas which would allow a satisfactory rounding out of the whole conceptual framework that is involved. The first of these deals with the question of a complete system of invariants. In §11.7, we shall prove that the (coefficients of the) two fundamental forms furnish such a system. The second line of ideas deals with the intrinsic geometry of a surface. We have met a number of quantities that were at first defined in terms of the imbedding of the surface, but which subsequently turned out to be functions of the g_{ik} only, and thus quantities belonging to the intrinsic geometry of the surface. Such was the case, for instance, of the geodesic curvature, or of geodesic parallelism, or of the Gaussian curvature. This state of affairs is unsatisfactory, at least in principle. One would like to see a development of the intrinsic geometry of surfaces based entirely on the g_{ik} and avoiding any dependence on the imbedding of the surface in space, so that all intrinsic geometric quantities can be introduced in an intrinsic way from the start. This kind of development of the theory of surfaces is one of the main concerns of Riemannian geometry, as presented in Chapter III.

Before entering into this development, however, we shall first deepen our grasp of the material studied so far by discussing some particularly important special classes of surfaces. By what has been said and by leaning on the model of the theory of space curves, we see that the following standard procedure for treating problems in the local theory of surfaces suggests itself: Formulate the assumptions of the given problem in terms of the vectors of the moving trihedron of the surface; express any derivatives of these vectors in terms of $\mathfrak{N}$, $\mathfrak{x}_1$, $\mathfrak{x}_2$ themselves by using the relations (5.25). The coefficients will then involve only the fundamental quantities g_{ik} and L_{ik} and their derivatives. By equating corresponding coefficients, obtain partial differential equations for the fundamental quantities and thus — at least in principle — obtain the solution of the problem. This purely schematic procedure often leads to the goal — not always, because it is not always possible to find explicit solutions of the differential equations. Thus, beyond the recipe-like scheme, what is often required is some inventiveness, for example, in such things as a skillful and appropriate choice of the system of parameters.

The reader might wish to look over the contents of the following section, as well as over the preceding ones, in the light of the general method just discussed.

6.2 Surfaces all of whose Points are Umbilics

Theorem 6.2.1. *If a piece of a surface consists entirely of umbilics and flat points, then it is a piece of a sphere or of the plane.*

Proof. At an umbilic, we have $L_{ik} = \lambda g_{ik}$; at a flat point, $\lambda = 0$. Multiplication by g^{ji} and summation over i yields $L^j{}_k = \lambda \delta^j{}_k$, so that Weingarten's equations (5.25) here assume the form

$$\mathfrak{N}_k = -\lambda \mathfrak{x}_k.$$

Differentiating with respect to u^i, we find

$$\mathfrak{N}_{ki} = -\mathfrak{x}_k \lambda_i - \mathfrak{x}_{ki} \lambda,$$

so that $0 = \mathfrak{N}_{ki} - \mathfrak{N}_{ik} = -\mathfrak{x}_k \lambda_i + \mathfrak{x}_i \lambda_k$, whence, for $i = 1$, $k = 2$, it follows that $\lambda_i = 0$, hence $\lambda = \text{const}$. Thus we can integrate Weingarten's equations explicitly, obtaining $\mathfrak{N} = -\lambda \mathfrak{x} + \mathfrak{d}$, where $\mathfrak{d}$ is a constant vector. Hence, if $\lambda = 0$, then the normal vector is constant and the surface is plane, while, if $\lambda \neq 0$, we obtain $\mathfrak{x} - \mathfrak{m} = \mathfrak{N}/\lambda$, $(\mathfrak{x} - \mathfrak{m})^2 = 1/\lambda^2$, which is the equation of a sphere.

6.3 Developables as the Only Surfaces of Vanishing Gaussian Curvature

We already know that on any developable surface, the Gaussian curvature is everywhere zero. This can be verified by direct calculation (see Exercise 3.7.12) or it can be deduced from the *theorema egregium* (Theorem 5.5.1), together with the applicability of developable surfaces on the plane (Exercise 3.7.3 and Theorem 3.7.2). We proceed to prove the converse.

Theorem 6.3.1. *If $K = 0$ at all points of a surface, then the surface is a developable surface.*[1]

Proof. The surface cannot contain umbilics but may contain flat points. By Theorem 6.2.1, any open portion that consists of flat points only is planar. We may, therefore, first confine our study to portions free of flat points and then extend the result, by a passage to the limit, to the flat points, if any, since these must be points of accumulation of nonflat points. On a portion without flat points, we can use the lines of curvature to serve as the parametric curves, so that we have $g_{12} = L_{12} = 0$ and, further, since $K = 0$, say $L_{22} = 0$ but $L_{11} \neq 0$ (as there are no flat points). It then follows from the second of Weingarten's equations that $\mathfrak{N}_2 = 0$. Next, we prove that

$$\mathfrak{x}_{22} = \lambda \mathfrak{x}_2. \tag{6.1}$$

For, since $L_{22} = 0$, it follows from Gauss' equation that $\mathfrak{x}_{22} = \Gamma^r_{22}\, \mathfrak{x}_r$; also, the Mainardi-Codazzi equation $C_{221} = 0$ implies that $\Gamma^1_{22}\, L_{11} = 0$, hence that $\Gamma^1_{22} = 0$, so that (6.1) is established, with $\lambda = \Gamma^2_{22}$. But (6.1) implies that the u^2-curves are straight lines along which the normal vector is constant, because $\mathfrak{N}_2 = 0$. Thus the surface portion is developable. For reasons of continuity it follows that through the flat points also there are straight lines that are contained in the surface and along each of which the normal vector remains constant.

The theorem just proved, in combination with the *theorema egregium*, leads to the following further result:

Theorem 6.3.2. *Each of the following properties of a surface implies the two others:*

---

[1] For a detailed discussion of this theorem, see S. S. Chern and R. K. Lashof, The total curvature of immersed manifolds, *Amer. J. Math.* **79** (1957), 308–318. See also L. Nirenberg, Elementary remarks on surfaces with curvature of fixed sign, *in* "Differential Geometry," Proceedings of Symposia in Pure Mathematics, Vol. III, pp. 181–185. Amer. Math. Soc., **1961**.

(a) *The surface is a developable.*

(b) *The surface consists entirely of parabolic points (and flat points), thus of points where* $K = 0$.

(c) *The surface is applicable on the plane, its intrinsic geometry is therefore the same as that of the plane.*

Exercise 6.3.3. *Prove*: Except for the plane, there does not exist any surface of vanishing mean curvature H that is applicable on the plane.

Exercise 6.3.4. *Prove*: The Gaussian curvature K of a ruled surface is never positive.

6.4 Bending of a Surface of Revolution

Let a surface of revolution be given by the equations $r = r(u^1)$, $x_3 = x_3(u^1)$ of its meridian, where u^1 is the arc length of the meridian. The surface itself is then given by

$$\mathfrak{x}(u^1) = \{r(u^1) \cos u^2, \quad r(u^1) \sin u^2, \quad x_3(u^1)\}. \tag{6.2}$$

The fundamental quantities are then computed (cf. Exercises 3.3.1, 3.5.1) to be

$$\begin{aligned} &g_{11} = 1, && g_{12} = 0, && g_{22} = r^2; && g = r^2, \\ &L_{11} = r' x_3'' - r'' x_3', && L_{12} = 0, && L_{22} = r x_3'; && L = r x_3'(r' x_3'' - r'' x_3'). \end{aligned} \tag{6.3}$$

The Gaussian curvature K depends on u^1 only; we have

$$K = K(u^1) = \frac{L}{g} = \frac{x_3'}{r}(r' x_3'' - r'' x_3'). \tag{6.4}$$

Since u^1 is arc length, we have $x_3'^2 + r'^2 = 1$, whence, by differentiation,

$$r' r'' + x_3' x_3'' = 0, \tag{6.5}$$

so that we can eliminate x_3'' in (6.4), obtaining

$$K(u^1) = -\frac{r''}{r}(r'^2 + x_3'^2) = -\frac{r''}{r}. \tag{6.6}$$

Theorem 6.4.1. *The Gaussian curvature* $K = K(u^1)$ *and the radius* $r = r(u^1)$ *of a circle of latitude of a surface of revolution satisfy the differential equation* $r'' + Kr = 0$.

If $K = K(u^1)$ is an assigned function, there always exist surfaces of revolution for which the Gaussian curvature along the meridian, with u^1 as arc length, is precisely $K(u^1)$, and all of these surfaces are given by

$$r'' + Kr = 0, \qquad x_3 = \int^{u^1} \sqrt{1 - r'^2(x)}\, dx.$$

(The differential equation for r has a one-parameter infinity of solutions, since one of the constants of integration only serves to fix the initial value of the parameter. The constant of integration in the integral for x_3 is not essential, as it means only a translation in the x_3-direction.)

Let us investigate the surfaces of revolution with $K = \text{const}$. For $K = 0$, Theorem 6.4.1 contains a new proof of the result of Exercise 3.5.1, which called for the determination of all surfaces of revolution applicable on the plane. The cases $K \neq 0$ are of greater interest. By a similarity transformation, not involving any changes that are of any significance here, we can always arrive at $K = \pm 1$.

Assume first that $K = 1$. In this case, all the solutions of (6.6) are given by

$$r(u^1) = C \cdot \cos (u^1 + \underset{0}{u}), \tag{6.7}$$

where we can always arrange for $\underset{0}{u} = 0$ by suitably fixing the initial point for the measurement of arc length. For the corresponding x_3,

$$x_3 = \int^{u^1} \sqrt{1 - C^2 \sin^2 x}\, dx. \tag{6.8}$$

For $C = 1$, the meridian is the unit circle; the corresponding surface of revolution is the unit sphere. For $C \neq 1$, (6.8) is an elliptic integral. Corresponding meridians are shown in Fig. 23. By (6.3), the element of arc length on a surface of revolution (6.7), (6.8) is given by

$$ds^2 = (du^1)^2 + C^2 \cos^2 u^1\, (du^2)^2;$$

since we can pass to new parameters $\bar{u}^1 = u^1$, $\bar{u}^2 = Cu^2$, we see that all of these surfaces are isometric to each other and to the unit sphere. They can be obtained from the unit sphere, slit along one of its meridians, by a continuous family of bending transformations. In this connection, we note that in the case $C > 1$, (6.8) shows that x_3 is not real for $|\sin u^1| > 1/C$.

Hence the surfaces of revolution for $C > 1$ are obtained by bending transformations from zones of the unit sphere bounded by the circles of latitude

$$u^1 = \pm \arcsin(1/C).$$

For these values of u^1, the surfaces obtained by the bending process have horizontal tangent planes at the corresponding points.

For $K = -1$, the general solution of (6.6) is

$$r(u^1) = C_1 \cosh u^1 + C_2 \sinh u^1.$$

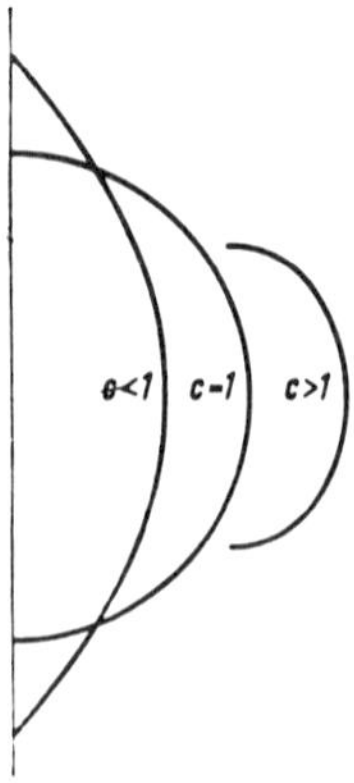

FIG. 23. Meridians of surfaces of revolution with $K = 1$.

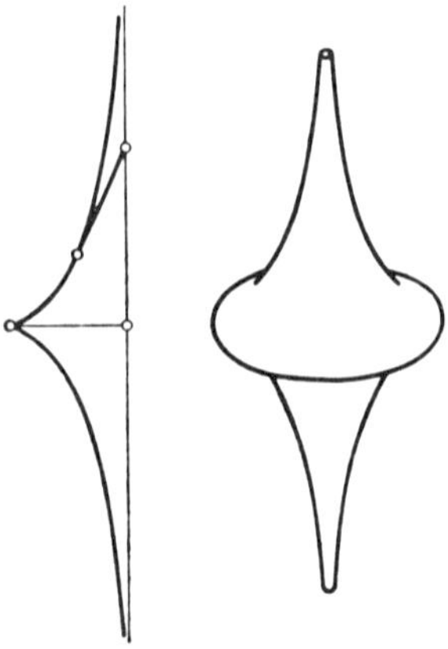

FIG. 24. Tractrix and pseudosphere.

The most interesting one of these surfaces of revolution is the *pseudosphere*, whose meridian is given by

$$r = e^{-u^1}, \qquad x_3 = \int^{u^1} \sqrt{1 - e^{-2x}}\, dx$$

and is called a *tractrix* or drag-curve, because the segment of its tangent between the point of contact and the x_3-axis is of constant length (here, equal to 1). See Fig. 24.

6.5 Bending Transformations of General Helicoids

The surfaces of revolution are the special cases $h = 0$ of the *general helicoids*; these are generated by a screw motion about the x_3-axis, of pitch $2\pi h$, of a "meridian curve" $r = \rho(\bar{u}^1)$, $x_3 = \xi_3(\bar{u}^1)$, and are, therefore, given by

$$\mathfrak{x}(\bar{u}^i) = \{\rho(\bar{u}^1) \cos \bar{u}^2;\quad \rho(\bar{u}^1) \sin \bar{u}^2;\quad \xi_3(\bar{u}^1) + h\,\bar{u}^2\}.$$

Without loss of generality, we may assume that u^1 is the arc length of the generating curve, hence that $\rho'^2 + \xi_3'^2 = 1$. We find

$$\bar{g}_{11} = 1, \qquad \bar{g}_{12} = h\xi_3', \qquad \bar{g}_{22} = \rho^2 + h^2,$$

so that

$$ds^2 = (d\bar{u}^1)^2 + 2h\xi_3'\, d\bar{u}^1\, d\bar{u}^2 + (\rho^2 + h^2)\,(d\bar{u}^2)^2.$$

This can be rewritten in the form

$$ds^2 = \left(1 - \frac{h^2\, \xi_3'^2}{\rho^2 + h^2}\right)(d\bar{u}^1)^2 + (\rho^2 + h^2)\left(d\bar{u}^2 + \frac{h\xi_3'}{\rho^2 + h^2}\, d\bar{u}^1\right)^2,$$

so that we can rewrite the element of arc, after an (admissible) parameter transformation, as follows:

$$du^1 = d\bar{u}^1 \sqrt{1 - \frac{h^2\, \xi_3'^2}{\rho^2 + h^2}}$$

$$du^2 = d\bar{u}^1 \frac{h\xi_3'}{\rho^2 + h^2} + d\bar{u}^2$$

$$ds^2 = (du^1)^2 + g_{22}\,(du^2)^2 \qquad \text{with} \qquad g_{22} = \rho^2 + h^2.$$

Here, g_{22} depends only on $\bar{u}^1$. In turn, the transformation formulas show that u^1 is given in terms of $\bar{u}^1$ only, not involving $\bar{u}^2$, so that their solution for $\bar{u}^1$ gives $\bar{u}^1 = \bar{u}^1(u^1)$; hence, g_{22} depends only on u^1 and not on u^2. Therefore, the element of arc is of the same form as it is for surfaces of revolution [see (6.3)] and we have thus proved the following result:

Theorem 6.5.1. (Bour, 1862). *A general helicoid is always applicable on a surface of revolution, with circles of latitude of radii* $r = r(u^1) = \sqrt{\rho^2(\bar{u}^1(u^1)) + h^2}$. *Under such an application, the trajectories* $\bar{u}^1 =$ const. *of the screw motion are mapped onto the circles of latitude* ($u^1 =$ const.) *of the surface of revolution.*

Exercise 6.5.2. Determine all surfaces of revolution whose mean curvature H vanishes identically. [Their meridians are catenaries (see Fig. 25),

$$x_3 = C_1 \operatorname{ar}\cosh(r/C_1) + C_2,$$

and the corresponding surfaces of revolution are *catenoids*.]

Exercise 6.5.3. *Prove*: The only general helicoids with $h \neq 0$ whose meridians are perpendicular to the trajectories of the screw motion are the winding surfaces, generated by the screw motion of a straight line perpendicular to the axis of this motion. For these surfaces, $H = 0$ holds.

Exercise 6.5.4. *Prove*: The surfaces of revolution on which the winding surfaces are applicable (in accordance with Theorem 6.5.1), are the catenoids (of Exercise 6.5.2).

Exercise 6.5.5. *Prove* that the quantities H and K do not constitute a complete system of invariants. (By Exercises 6.5.2 and 6.5.3, the winding surfaces and the catenoids have vanishing mean curvature H; by Exercise 6.5.4 and the *theorema egregium,* their Gaussian curvatures K are equal at corresponding points; but the

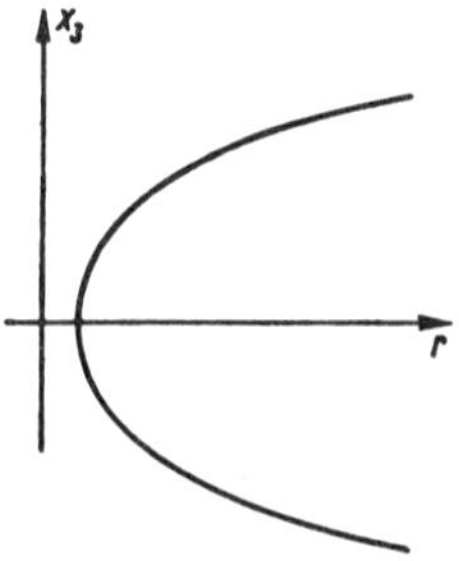

FIG. 25. Catenary.

surfaces cannot be mapped onto each other by a rigid motion, since one of them is a ruled surface, while the other one is not. — The same argument even proves that H and g_{ik} do not suffice for a complete system of invariants.)

6.6 The Surfaces of Vanishing Mean Curvature as Minimal Surfaces

The surfaces with $H = 0$ possess a remarkable extremal property, on account of which they are called *minimal surfaces.*

Theorem 6.6.1. *Let $\mathfrak{x} = \mathfrak{x}(u^i)$ be a portion of a surface and let the closed curve R be the boundary of this portion. If the area of this piece of surface is smaller than, or at most equal to, the area of every (sufficiently near) other piece of surface bounded by the same closed curve R, then $H = 0$ holds on $\mathfrak{x} = \mathfrak{x}(u^i)$.*

Proof. Let us consider the following family of neighboring surfaces:

$$\bar{\mathfrak{x}}(u^i) = \mathfrak{x}(u^i) + \varepsilon v(u^i)\mathfrak{N}(u^i).$$

It follows that

$$\bar{\mathfrak{x}}_i = \mathfrak{x}_i + \varepsilon v_{,i}\,\mathfrak{N} + \varepsilon v\mathfrak{N}_i = \mathfrak{x}_k(\delta_i^k - \varepsilon v L_i^k) + \varepsilon v_{,i}\,\mathfrak{N},$$

by Weingarten's equation (5.23). This implies further that

$$\bar{g}_{ij} = g_{ij} - 2\varepsilon v L_{ij} + \varepsilon^2(\ldots)$$

and, with $g = \det(g_{ik})$, we obtain on account of $H = \frac{1}{2} g^{lm} L_{lm}$ that

$$\bar{g} = g(1 - 4\varepsilon v H) + \varepsilon^2(\ldots)$$

and finally that

$$\left.\frac{\partial \sqrt{\bar{g}}}{\partial \varepsilon}\right|_{\varepsilon = 0} = -2v\sqrt{g}H.$$

If the area is to be a minimum at $\varepsilon = 0$, then we must have

$$\left.\frac{\partial}{\partial \varepsilon} \iint \sqrt{\bar{g}}\, du^1\, du^2 \right|_{\varepsilon = 0} = - \iint 2vH \sqrt{g}\, du^1\, du^2 = 0,$$

for all admissible functions $v(u^i)$. From this it follows, by the fundamental lemma of the Calculus of Variations (see Appendix II.3), that $H = 0$, as we set out to prove.

These surfaces are remarkable in many ways. We shall return to them later, in §13.8.

We have met examples of minimal surfaces, other than the plane, in the last set of exercises, namely the winding surfaces and the catenoids. It is easy to produce models of minimal surfaces: Form the given (closed) boundary curve out of wire, immerse it in a soap solution, and pull it out; the resulting soap film will be in the shape of a minimal surface (to be sure, only if the slight effect of gravity is discounted). In 1866, the physicist Plateau proposed the question as to whether there is for every closed curve a minimal surface bounded by that curve. This "problem of Plateau," which leads to complicated analytic proofs, was answered in the positive by J. Douglas (1931) and T. Radó (1930).

There is a large literature devoted to these surfaces, whose interesting history can be gleaned, e.g., from the relevant papers of H. A. Schwarz.[1] More recent references will be found in R. Courant: "Dirichlet's Principle, Conformal Mapping and Minimal Surfaces." New York, 1950.

6.7 Cornices

We shall now investigate another special class of surfaces which, like the class of general helicoids, also includes the surfaces of revolution. On every surface of revolution, the meridians are a family of geodesics that are, at the same time, plane curves. On a sphere, all geodesics are plane curves. We show first that the last property characterizes spheres.

Theorem 6.7.1. *If all geodesics of a surface are plane curves then the surface is a piece of a sphere or of the plane.*

---

[1] "Collected Mathematical Works," Vol. I. Berlin, 1890.

Proof. We first prove the following:

Lemma. *A plane geodesic, other than a straight line, must be a line of curvature.*

(*Proof of the lemma.* The plane of the geodesic is also its osculating plane and, therefore, contains the surface normals. Hence, these normals generate a developable surface and this makes the geodesic a line of curvature, by Exercise 5.3.3.)

Now to prove the theorem, consider any point of the surface other than a flat point. There is at least one direction through this point such that, for the geodesic in this direction, $\mathfrak{x}''$ does not lie along $\mathfrak{x}'$. But $\mathfrak{N}'$ must then lie in this plane, hence

$$0 = \mathfrak{N}' \times \mathfrak{x}' = \mathfrak{N}_1 \times \mathfrak{x}_1(u'^1)^2 + (\mathfrak{N}_1 \times \mathfrak{x}_2 + \mathfrak{N}_2 \times \mathfrak{x}_1)u'^1 u'^2 + \mathfrak{N}_2 \times \mathfrak{x}_2(u'^2)^2.$$

The same holds true, not only for the direction just considered, but for infinitely many nearby directions as well, whence it follows that

$$\mathfrak{N}_1 \times \mathfrak{x}_1 = \mathfrak{N}_2 \times \mathfrak{x}_2 = \mathfrak{N}_1 \times \mathfrak{x}_2 + \mathfrak{N}_2 \times \mathfrak{x}_1 = 0.$$

The first two relations imply $\mathfrak{N}_i = \lambda_i \mathfrak{x}_i$ and substitution into the last relation yields $\lambda_1 = \lambda_2$. Therefore the point in question is an umbilic. Thus the surface consists entirely of umbilics and flat points, whence the assertion to be proved follows by Theorem 6.2.1.

We turn next to the problem of determining all surfaces on which there is at least one family of plane geodesics. This class certainly includes all ruled surfaces, since straight lines are plane geodesics. For the case that not all curves of the given family are straight lines, we confine our attention to open portions of the surface in which no curve of the family is a straight line and which we also assume to be free of umbilics — this last only excludes spheres. Under these conditions, we can use the lines of curvature to serve as the parametric net. By the lemma under Theorem 6.7.1, the curves of the given family are lines of curvature and thus we can take them to be, say, the u^1-curves. The parametric net being also geodesic, we may assume $g_{11} = 1$. The equations of Mainardi and Codazzi for this case (see Exercise 5.5.6) imply that $L_{11,\,2} = 0$ or $L_{11} = k(u^1)$. Since curvature and normal curvature are the same along geodesics, $k(u^1)$ is the curvature of the u^1-curve, and the torsion of this (plane) curve vanishes identically. It follows, therefore, from the fundamental theorem of the theory of space curves that all of the u^1-curves are mutually congruent. Now, a surface is called a *cornice* surface if it intersects orthogonally every plane of a family of planes, the curves of intersection all being congruent. (Such a surface can be generated kinematically: Draw the u^1-curve in one of these planes; then roll the plane off on the developable surface that envelops the family of planes.) We have therefore proved the following result:

Theorem 6.7.2. *A surface that contains a family of plane geodesics is either a ruled surface or a cornice.*

The cornices include, besides the surfaces of revolution, another special class of surfaces called *tubular surfaces* (or *canal surfaces of constant radius*); the above-mentioned family of planes is, for these, the family of normal planes of a fixed space curve $\mathfrak{x}(s)$, while the u^1-curves are circles of a fixed radius r and with their centers on $\mathfrak{x}(s)$. The parametric representation of a tubular surface, if we set $s = u^2$, is

$$\mathfrak{x}(u^i) = \mathfrak{x}(u^2) + r(\mathfrak{v}_2(u^2)\cos u^1 + \mathfrak{v}_3(u^2)\sin u^1).$$

Exercise 6.7.3. *Prove*: There does not exist any piece of surface free of umbilics and possessing two families of nonrectilinear plane geodesics. From this, the following sharpening of Theorem 6.7.1 is obtained as a corollary: A surface with two families of nonrectilinear plane geodesics is spherical [D. Laugwitz, *Arch. Math.* 8 (1957), 472–476].

Exercise 6.7.4. *Prove*, using Frenet's formulas for $\mathfrak{x}(s)$, that the following hold for tubular surfaces:

$$\begin{aligned}
\mathfrak{x}_1 &= -\mathfrak{v}_2 r \sin u^1 + \mathfrak{v}_3 r \cos u^1 \\
\mathfrak{x}_2 &= \mathfrak{v}_1(1 - \kappa r \cos u^1) - \mathfrak{v}_2 \tau r \sin u^1 + \mathfrak{v}_3 \tau r \cos u^1 \\
g &= (\mathfrak{x}_1 \times \mathfrak{x}_2)^2 = r^2(1 - \kappa r \cos u^1)^2.
\end{aligned}$$

These, therefore, represent a surface if $r < 1/\kappa$. *Show* that the Gaussian curvature K and the curvature κ of the curve are related by

$$K = -\kappa \cos u^1 / \sqrt{g}.$$

(In this connection, see also V, §16.4.)

6.8 Triply Orthogonal Systems of Surfaces

Let $\mathfrak{y}(u^1, u^2, u^3)$ be of class C^2, assume that the $\mathfrak{y}_i = \partial\mathfrak{y}/\partial u^i$ are linearly independent, and assume finally that $\mathfrak{y}_i\,\mathfrak{y}_k = 0$ for $i \neq k$. Then the surfaces $u^i =$ const. constitute three mutually orthogonal families of surfaces, a so-called *triply orthogonal system*.

Theorem 6.8.1. (Dupin, 1813). *The curves of intersection of the surfaces of a triply orthogonal system are lines of curvature on these surfaces.*

Proof. Differentiation of $\mathfrak{y}_i\,\mathfrak{y}_k = 0$ $(i \neq k)$ yields

$$\mathfrak{y}_{12}\,\mathfrak{y}_3 + \mathfrak{y}_1\,\mathfrak{y}_{23} = \mathfrak{y}_{23}\,\mathfrak{y}_1 + \mathfrak{y}_2\,\mathfrak{y}_{13} = \mathfrak{y}_{21}\,\mathfrak{y}_3 + \mathfrak{y}_2\,\mathfrak{y}_{31} = 0;$$

hence,

$$\mathfrak{y}_{12}\,\mathfrak{y}_3 = \mathfrak{y}_{23}\,\mathfrak{y}_1 = \mathfrak{y}_{31}\,\mathfrak{y}_2 = 0.$$

Since the three vectors $\mathfrak{y}_1$, $\mathfrak{y}_2$, $\mathfrak{y}_{12}$ all are orthogonal to $\mathfrak{y}_3$, they are linearly dependent,

$$\mathfrak{y}_1\,\mathfrak{y}_2\,\mathfrak{y}_{12} = 0,$$

which implies $L_{12} = 0$, in addition to which we have $\mathfrak{y}_1\,\mathfrak{y}_2 = g_{12} = 0$ for the surfaces $u^3 = \text{const}$. Therefore, the parametric curves are lines of curvature and this applies similarly to the surfaces of the other two families.

The following result is a partial converse:

Theorem 6.8.2. (Darboux's Imbedding Theorem). *If the parametric curves of a surface $\mathfrak{x} = \mathfrak{x}(u^i)$ are its lines of curvature, then for sufficiently small u^3,*

$$\mathfrak{y}(u^1, u^2, u^3) = \mathfrak{x}(u^1, u^2) + u^3\,\mathfrak{N}$$

is a triply orthogonal system.

Proof. We know not only that $g_{12} = L_{12} = 0$, but also, by Rodrigues formulas (see Exercise 5.3.2), that $\mathfrak{N}_i$ is proportional to $\mathfrak{x}_i$. Hence $\mathfrak{y}_1 = \mathfrak{x}_1 + u^3\,\mathfrak{N}_1$, $\mathfrak{y}_2 = \mathfrak{x}_2 + u^3\,\mathfrak{N}_2$, $\mathfrak{y}_3 = \mathfrak{N}$ imply that

$$\mathfrak{y}_1\,\mathfrak{y}_2 = \mathfrak{y}_2\,\mathfrak{y}_3 = \mathfrak{y}_3\,\mathfrak{y}_1 = 0,$$

which proves the asserted orthogonality. Since $\mathfrak{x}_1 \times \mathfrak{x}_2 \neq 0$, we also have, for sufficiently small u^3, that

$$\mathfrak{y}_1 \times \mathfrak{y}_2 \neq 0,$$

so that we are indeed dealing with surfaces.

Exercise 6.8.3. *Confocal Quadrics* (Second-Order Surfaces). This designation applies to the surfaces of a family

$$g(t) \underset{\text{def}}{=} \frac{x_1^2}{t - a_1} + \frac{x_2^2}{t - a_2} + \frac{x_3^2}{t - a_3} = 1.$$

Show that if $a_1 > a_2 > a_3 > 0$, there pass through every point of space with $x_k \neq 0$ three surfaces of the family mutually orthogonal at the point. [*Hint.* For $\mathfrak{x}$ fixed,

$g(t) = 1$ is a cubic equation in t. Since $f(t) = g(t) - 1$ jumps from $-\infty$ to $+\infty$ at each a_k and is continuous elsewhere, the equation has three real solutions, with $t_1 > a_1 > t_2 > a_2 > t_3 > a_3$. Corresponding to t_1 there is an ellipsoid through $\mathfrak{x}$, to t_2 a hyperboloid of one sheet, to t_3 a hyperboloid of two sheets. The t_k can be introduced as new coordinates. The surface normal for the surface $t_j =$ const. has the direction $b^i{}_{(j)}\, \mathfrak{e}_i$, where $b^i{}_{(j)} = x_i/(t_j - a_i)$. The inner product of the normal vectors to the surfaces t_j and t_k is

$$\sum_{i=1}^{3} \frac{x_i^2}{(t_j - a_i)\,(t_k - a_i)},$$

which equals zero, as follows from $g(t_j) - g(t_k) = 0$.]

Hence the lines of curvature of these quadric surfaces are certain space curves of degree four (quartic curves) which one obtains by intersecting with confocal surfaces. For a very graphic discussion of these families of surfaces, see D. Hilbert and St. Cohn-Vossen, "Anschauliche Geometrie." Berlin, 1932 (English translation: "Geometry and the Imagination." New York, 1952).

Exercise 6.8.4. Discuss the case in which one surface in a triply orthogonal system is a sphere. (Every curve of intersection is a line of curvature; the surfaces of the other two families are cones, the surfaces of the first family all are spheres concentric with the given sphere.)

CHAPTER III

Tensor Calculus and Riemannian Geometry

§ 7. The Concept of a Differentiable Manifold

7.1 Preliminary Remarks

Before continuing with the development of differential geometry proper, we shall have to deal with some very basic, though more formal problems. We recall that our next geometric goal will be a new treatment, independent of imbedding, of the intrinsic geometry of surfaces, based entirely on the definition of measurement of lengths within the point set itself. The first to do this was Riemann, in his famous inaugural address at Göttingen in 1854, delivered in the presence of the elder Gauss.

To this end, we must have a concept of a geometric manifold that is independent of its imbedding in Euclidean space. It will turn out that the

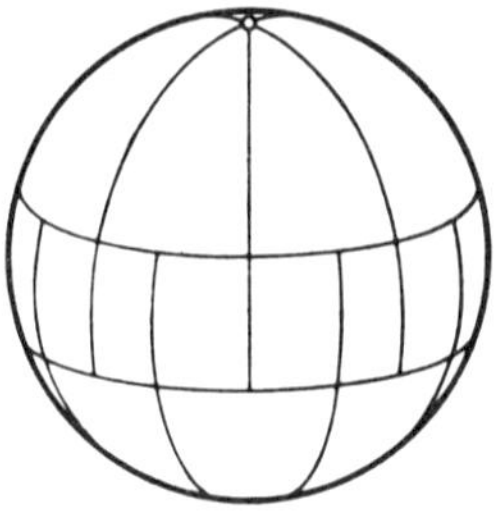

FIG. 26. Coordinates on a sphere.

necessary definitions can be formulated so as to apply, without additional apparatus, to dimensions $n > 2$. Our manifolds will therefore include the Euclidean n-dimensional spaces as special cases.

Our definition of the concept of surface in §3.1 will prove too narrow for our later purposes in another respect as well. That definition only applies to portions of surfaces, but not necessarily to surfaces in their full extension. Thus for instance, the complete surface of a sphere does not come under our old definition, since it is impossible to map the full sphere one-to-one and continuously on a single region of the plane of parameters. To be sure, one can get around this, in the special case just given, by separately mapping two spherical caps onto parameter regions, say, the one cap extending from the north pole to a southern circle of latitude, the other, from the south pole to a northern latitude (see Fig. 26). For the equatorial strip on which the two caps overlap, one will have to require that the corresponding parametrizations be such that the transition from one to the other is an allowable parameter transformation.

We recall also the importance of the tangent plane for the intrinsic geometry of surfaces. Geodesic parallel displacement, which we showed to be an intrinsic-geometric concept, deals with tangent vectors, which are vectors that (usually) point outside, into the imbedding space. Now we shall no longer have an imbedding space at our disposal; however, we shall construct, at each point of an n-dimensional manifold, an n-dimensional vector space from the "tangents" of differentiable curves in the manifold, and this vector space will correspond to the tangent plane.

Calculations involving indices (superscripts and subscripts) and the summation convention, having already shown their practical value, will find a deeper justification and a further development in tensor algebra. The equations for derivatives of Gauss and Weingarten, so basic in the theory of surfaces, will no longer be at our disposal because they involve the imbedding in space. We shall discover a substitute for them in tensor analysis, which permits the differentiation of quantities within the manifold itself.

7.2 The Definition of Differentiable Manifold

We assume as given a *Hausdorff space*, that is, a set of elements called *points* such that in that set, certain subsets — called *open sets* — are distinguished among the totality of all subsets, and such that the following four propositions hold true:

1. The entire space is one of the open sets, and (for formal reasons) so is the empty set.

2. The intersection of two open sets is open. (Since the empty set is open by 1, proposition 2 is meaningful also in case the intersection is empty.)

3. The union of any collection of open sets is an open set.

4. For any two distinct points P, Q of the space there are open sets U_P, U_Q ("neighborhoods") such that P is an element of U_P, Q is an element of U_Q, and the intersection of U_P and U_Q is empty.

(The reader should verify that every open subset of Euclidean space or of the Euclidean plane is a Hausdorff space if "open set" is given the usual definition in these cases.)

A sequence P_n of points of a Hausdorff space is said to be *convergent to* P (in symbols, $P_n \Rightarrow P$) if every open set that contains P also contains almost all P_n. The separation property 4 ensures that any given sequence converges to at most one point. A mapping $P \rightarrow \bar{P}$ of one Hausdorff space into another is called *continuous* if $P_n \Rightarrow P$ always implies $\bar{P}_n \Rightarrow \bar{P}$. We are now ready for our definition.

Definition. *By an n-dimensional differentiable manifold of class* C^k ($k = 1, 2, \ldots,$ *or* $k = \infty$) *is meant a Hausdorff space with a system* $\mathfrak{U} = \{U_\mu\}$ *of open sets and a system of mappings* f_μ *satisfying the following conditions* (a), (b), (c):

(a) *Every point is contained in at least one of the* U_μ *(so that the space is covered by the* U_μ*).*

(b) f_μ *is a one-to-one and bicontinuous mapping (i.e., a homeomorphism) from* U_μ *onto an open subset* O_μ *of n-dimensional Euclidean space,* $\underset{\mu}{x}^i(P) = f_\mu(P)$ *for all* P *of* U_μ.

(c) *If a point* P *(and therefore, an entire neighborhood of* P*) is contained in two sets* U_μ *and* U_ν*, there is thereby induced a mapping*

$$\underset{\mu}{x}^j = \underset{\mu}{x}^j(\underset{\nu}{x}^i)$$

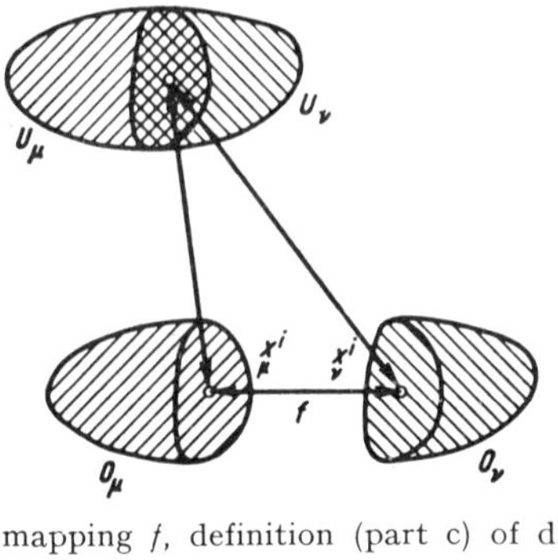

FIG. 27. The induced mapping f, definition (part c) of differentiable manifold.

in Euclidean n-space (see Fig. 27). *This mapping is assumed to be of class* C^k *and to have a functional determinant satisfying*

$$\partial(\underset{\mu}{x}^j)/\partial(\underset{\nu}{x}^i) \neq 0.$$

Addendum to the Definition. If the induced mapping is analytic, i. e., if the functions representing it have power series expansions, then the manifold is called *analytic*, or *of class* C^ω.

The mappings f_μ furnish, as it were, a first, or original, system of coordinates for the points of the manifold. It will be advantageous to make allowance for changes to other coordinate systems.

In this sense, let us assume that the given manifold admits of a second covering $\bar{U}_\lambda$ and corresponding functions $\bar{f}_\lambda$: $\bar{U}_\lambda \to \bar{O}_\lambda$, likewise satisfying properties (a), (b), (c). Then the change from the original system to this new one is called an *admissible* (or *allowable*) coordinate transformation provided that, for every intersection of a U_μ and a $\bar{U}_\lambda$, the induced mapping $\bar{x}^k_\lambda = \bar{x}^k_\lambda(x^i_\mu)$ between the corresponding parts of O_μ and $\bar{O}_\lambda$ is of class C^k at least (or is analytic, if the given manifold is of class C^ω), and has a nonvanishing functional determinant.

So far, a differentiable manifold carries a topological structure but has not yet been endowed with a differential-geometric structure. Of geometric properties to be introduced we shall ask that they be independent of the choice of a coordinate system, just as we did in the theory of surfaces. We shall now take up the differentiability properties of curves, as being such geometric properties.

7.3 The Tangent Space

A continuous map $P(t)$ of a Euclidean segment $a < t < b$ into a manifold is a *curve* in the manifold. If the manifold is of class C^k and if $m \leq k$, a curve is said to be of class C^m if the $x^i(t)$ are of class C^m, in some coordinate system. This definition makes sense, as that of a geometric property, because

$$\bar{x}^j(t) = \bar{x}^j(x^i(t))$$

is then of class C^m in every other admissible coordinate system, as follows from the chain rule. A curve is called *smooth* if it is of class C^1 at least, and a curve pieced together from a finite number of smooth curves is called *piecewise smooth*.

Let P be a given point of a manifold. Two smooth curves passing through P, given in coordinates by $x^i(t)$ and $y^i(\tau)$, are said to have the same tangent vector $\mathfrak{t}$ at P if $dx^i/dt = dy^i/d\tau$ holds at P. This definition is also independent of the coordinate system used, as can again be seen by the chain rule. It allows the setting up of an equivalence relation, independent of the coordinate system, among all smooth curves passing through P: Two such curves are equivalent if they have the same tangent vector $\mathfrak{t}$ at P. (The reader should verify that the three properties of an equivalence relation — symmetry, reflexivity, and transitivity — are indeed satisfied here.)

Now we can define *tangent vectors* themselves as being the equivalence classes determined by the above equivalence relation. We shall show that the tangent vectors, thus defined, constitute an n-dimensional vector space over the real numbers. To begin with, if $x^i(t)$ has the tangent vector $\mathfrak{t}$ at P, then, for the curve $z^i(t) = x^i(c \cdot t)$, we have

$$dz^i/dt = c \cdot dx^i/dt$$

at P and, if the curves $x^i(t)$ range over an equivalence class, then so do the curves $x^i(c \cdot t)$. Let us denote this last class by $c \cdot \mathfrak{t}$. We must still define an addition for tangent vectors. This can again be done in a special coordinate system, provided the sum is then shown to be independent of the special coordinate system that was used. Let $x^i(t)$, $x_1^i(t)$ range over all curves of an equivalence class or tangent vector $\mathfrak{t}$ at P and let $y^i(t)$, $y_1^i(t)$ range over all curves of a second such class or tangent vector $\mathfrak{u}$. Then all curves $x^i(t) + y^i(t)$ belong to one class, for if $\dot{x}^i = \dot{x}_1^i$, $\dot{y}^i = \dot{y}_1^i$ hold true at P, then so does $d(x^i + y^i)/dt = d(x_1^i + y_1^i)/dt$. We denote this class by $\mathfrak{t} + \mathfrak{u}$. Its independence of the coordinate system follows once again from the chain rule.

The tangent vectors, under the above multiplication by real numbers and the above addition, constitute a vector space, as can be verified directly by checking the validity of the vector space axioms (for these, consult any textbook on linear algebra) in a fixed coordinate system. This vector space is n-dimensional, since

$$x^i(t) = t\delta_k^i \qquad \left(\delta_k^i = \begin{matrix} 1 & \text{for} & i = k \\ 0 & \text{for} & i \neq k \end{matrix}\right)$$

are curves with the tangent vectors $\dot{x}^i = \delta_k^i$ and from these, we can obtain any other tangent vector as a linear combination, as $\dot{y}^i = a^k \delta_k^i$ with $a^k = \dot{y}^k$.

Thus we see that with every point of an n-dimensional differentiable manifold there is associated an n-dimensional vector space, the *tangent space* at that point. We shall need in the sequel some important algebraic facts concerning such vector spaces and we shall develop these in §8 below.

7.4 Connected Manifolds

A topological space is said to be *connected* if it is not the union of two nonempty disjoint open subsets. For our purposes, another (though related) definition will be more important: A manifold is *arcwise connected* if, for any two of its points P, Q, there is a continuous curve in the manifold that joins P and Q.

We can show that the curve can, in this case, always be chosen so as to be piecewise smooth. Assume that a fixed coordinate system has been chosen and let C be a continuous curve, corresponding to a parameter interval $a \leqq t \leqq b$. The intersection of any coordinate neighborhood with C corresponds to an open t-interval, and every point of the curve is covered by at least one coordinate neighborhood. Therefore, by the *Heine-Borel* covering theorem, a finite number of coordinate neighborhoods suffice to cover the curve. Choose a finite number of t_k, with $a = t_0 < t_1 < \ldots < t_n = b$, in such a way that, for each interval $t_k \leqq t \leqq t_{k+1}$, the corresponding piece of the curve is covered by a single one of the coordinate neighborhoods, with coordinates $x^i(t)$. By the *Weierstrass* approximation theorem, the continuous functions $x^i(t)$ can be approximated by polynomials $y^i(t)$, with $y^i(t_k) = x^i(t_k)$, $y^i(t_{k+1}) = x^i(t_{k+1})$, sufficiently closely to guarantee that the $y^i(t)$ still belong to the same coordinate neighborhood. Being polynomials, the $y^i(t)$ are analytic and are therefore of class C^k in any other (allowable) coordinate system if the manifold is of class C^k.

All differentiable manifolds to be considered in the sequel will be assumed to be arcwise connected and we shall make use of the fact that any two points of such a manifold can be joined by a piecewise smooth curve.

§ 8. Tensor Algebra

8.1 Change of Basis in a Vector Space

Let V be an n-dimensional vector space over the field of real numbers. The elements of V will be denoted by lower-case German letters. If $\mathfrak{n}_i$ $(i = 1, \ldots, n)$ is a linearly independent n-tuple of vectors of V, then every vector $\mathfrak{x}$ of V can be represented in one and only one way as a linear combination

$$\mathfrak{x} = x^i \mathfrak{n}_i \tag{8.1}$$

of the basis vectors $\mathfrak{n}_i$. If an n-tuple of vectors $\bar{\mathfrak{n}}_k$ constitutes another basis, then, by (8.1), the $\bar{\mathfrak{n}}_k$ can be represented in terms of the $\mathfrak{n}_i$, thus

$$\bar{\mathfrak{n}}_k = a_k{}^i \mathfrak{n}_i \tag{8.2}$$

and, conversely, the old basis vectors can be expressed in terms of the new ones, thus

$$\mathfrak{n}_j = \alpha_j{}^k \bar{\mathfrak{n}}_k. \tag{8.3}$$

Equations (8.2) and (8.3) jointly describe a change of basis. If (8.3) is substituted into (8.2), it follows from the linear independence of the vectors of any given basis that

$$\alpha_j{}^k a_k{}^i = \delta_j^i, \tag{8.4}$$

and substitution of (8.2) into (8.3) similarly yields

$$a_k{}^i \alpha_i{}^j = \delta_k^j. \tag{8.5}$$

Each of the two transformation matrices $(a_j^{\ k})$ and $(\alpha_k^{\ i})$ is of rank n, and the two matrices are inverses of each other.

Any given vector $\mathfrak{x}$ of V can be expressed in terms of either basis, and using

$$\mathfrak{x} = x^j\,\mathfrak{n}_j = x^j\,\alpha_j^{\ k}\,\bar{\mathfrak{n}}_k = \bar{x}^k\,\bar{\mathfrak{n}}_k = \bar{x}^k\,a_k^{\ j}\,\mathfrak{n}_j$$

we are led to the following transformation formulas for the vector components:

$$\bar{x}^k = \alpha_j^{\ k}\,x^j, \qquad x^j = a_k^{\ j}\,\bar{x}^k. \tag{8.6}$$

Example. Let $n = 2$, and let a change of basis be given by $\bar{\mathfrak{n}}_1 = \mathfrak{n}_1 + \mathfrak{n}_2$, $\bar{\mathfrak{n}}_2 = \mathfrak{n}_2$. Then the corresponding transformation for vector components is $\bar{x}^1 = x^1$, $\bar{x}^2 = -x^1 + x^2$.

One obtains instructive and less trivial examples if there is defined, over the vector space V, an inner product $\mathfrak{x} \cdot \mathfrak{y}$ with the usual properties, such that $\mathfrak{x} \cdot \mathfrak{x} = \mathfrak{x}^2$ equals the square of the Euclidean length of the vector $\mathfrak{x}$. For such spaces, special importance attaches to orthonormal bases $\mathfrak{n}_i$, i.e., to bases for which

$$\mathfrak{n}_i \cdot \mathfrak{n}_k = \delta_{ik}.$$

Exercise 8.1.1. *Prove* that a change of basis (8.2) transforms an orthonormal basis $\mathfrak{n}_i$ into another such basis $\bar{\mathfrak{n}}_k$ if and only if $a_i^{\ j} = \alpha_j^{\ i}$.

This exercise shows why, in the most common case – that of Euclidean spaces referred to rectangular Cartesian coordinates – it is not necessary to distinguish between the transformation that applies to the basis vectors and that which applies to vector components. For, in this case, where (as an exception) we write all indices as subscripts, it follows from (8.6) that if $\bar{\mathfrak{n}}_k = a_{ki}\,\mathfrak{n}_i$ then $\bar{x}_k = \alpha_{ik}\,x_i = a_{ki}\,x_i$, so that the two transformations just mentioned have the same matrix. In general, however, the matrix (α_{ki}) belonging to the transformation (8.6) of vector components is the transpose of the inverse (sometimes called the contragredient) of the matrix (a_{ik}) belonging to the change (8.2) of basis.

8.2 Linear Functionals, and the Dual Vector Space

A real-valued function $l(\mathfrak{x})$ defined on a vector space V is called a *linear functional* or *linear form* on V if

$$l(a\mathfrak{x} + b\mathfrak{y}) = a \cdot l(\mathfrak{x}) + b \cdot l(\mathfrak{y}) \tag{8.7}$$

holds for all vectors $\mathfrak{x}$, $\mathfrak{y}$ of V and for all real numbers a, b. It follows that

$$l(\mathfrak{x}) = l(x^i\,\mathfrak{n}_i) = x^i\,l(\mathfrak{n}_i). \tag{8.8}$$

Hence a linear functional is completely determined by the n numbers $l(\mathfrak{n}_i)$. Conversely, if l_i is any assigned n-tuple of real numbers, then the function

$$l(\mathfrak{x}) = x^i\,l_i$$

is a linear functional on V. Hence, the totality of linear functionals on V is an n-dimensional vector space, the addition of two linear functionals l, m and scalar multiplication by a real number a being defined in the natural manner as follows:

$$(l + m)(\mathfrak{x}) \underset{\text{def}}{=} l(\mathfrak{x}) + m(\mathfrak{x}),$$

$$(a \cdot l)(\mathfrak{x}) \underset{\text{def}}{=} a \cdot l(\mathfrak{x}).$$

This vector space of linear functionals on V is associated with V independently of the choice of a basis in V and is called the *dual* vector space of V. However, for any given basis $\mathfrak{n}_i$ of V, there is in the dual space a corresponding basis, consisting of the functionals that assume the value 1 at one of the $\mathfrak{n}_i$ and vanish at all remaining $\mathfrak{n}_k$. We write $\mathfrak{n}^i$ for this dual basis and we therefore have [writing simply $\mathfrak{n}^i\mathfrak{n}_k$ for $\mathfrak{n}^i(\mathfrak{n}_k)$] that

$$\mathfrak{n}^i\,\mathfrak{n}_k = \delta^i_k. \tag{8.9}$$

It is easy to verify explicitly that every vector l of the dual space has a unique representation as a linear combination of the $\mathfrak{n}^i$:

$$l = l(\mathfrak{n}_i)\mathfrak{n}^i.$$

This follows from (8.8) and (8.9), since $\mathfrak{n}^i(\mathfrak{x}) = x^i$, by (8.9).

We shall now investigate the laws of transformation for the dual basis $\mathfrak{n}^i$ and for the components l_i of a vector of the dual space if a new basis $\bar{\mathfrak{n}}_k$ is introduced in V by means of

$$\mathfrak{n}_i = \alpha_i{}^k\,\bar{\mathfrak{n}}_k, \qquad \bar{\mathfrak{n}}_k = a_k{}^i\,\mathfrak{n}_i. \tag{8.10}$$

Since $l_i = l(\mathfrak{n}_i)$, the linearity property (8.7) yields

$$l_i = \alpha_i{}^k\,\bar{l}_k, \qquad \bar{l}_k = l(\bar{\mathfrak{n}}_k) = a_k{}^i\,l_i. \tag{8.11}$$

Hence, the components of a dual vector are transformed in exactly the same way as the basis vectors of V, as a comparison of (8.11) with (8.10) shows. For this reason, one calls the vectors of the dual space *covariant* vectors:

Definition. *The vectors of V are called contravariant vectors; those of the dual space of V are called covariant vectors. The transformation of their components under a change of basis* (8.10) *is governed by* (8.6) *and* (8.11), *respectively.*

The term "contravariant vectors" stems from the fact that the components of such vectors are transformed contragrediently (i.e., by the transpose of the inverse matrix) as compared to the basis vectors themselves. The use of such an attribute as applied to the vectors of V is a useful device, for emphasizing the distinction between them and the covariant vectors. The transformation formulas for the two kinds of vectors also contain the information needed for their distinction in the position of the indices on the vector components: The components of contravariant vectors carry superscripts, those of covariant vectors, subscripts.

Exercise 8.2.1. *Prove* that under a change of basis (8.10) in V, the vectors of the dual basis are transformed like components of contravariant vectors.

Exercise 8.2.2. Under a change from one orthonormal basis to another (see the end of § 8.1), the distinction between the two kinds of transformation laws vanishes. In this case, one may as well write all indices as subscripts. What can be said about the transformation of the dual basis?

8.3 Tensors of the Second Order

Let U, W be two real vector spaces. A real-valued function $B(\mathfrak{x}, \mathfrak{X})$ defined for all $\mathfrak{x}$ of U and $\mathfrak{X}$ of W is called a *bilinear form* if, for every fixed $\mathfrak{X}$, it is a linear form on U, and for every fixed $\mathfrak{x}$, a linear form on W. We shall be interested mainly in the case where the choices of U and W are made from among a vector space V and its dual space V_1. We consider the three essentially different possibilities

$$\text{(a)}\quad U = W = V, \qquad \text{(b)}\quad U = V,\quad W = V_1, \qquad \text{(c)}\quad U = W = V_1.$$

The three corresponding types of bilinear forms will introduce us to the three types of tensors of the second order.

(a) If $b(\mathfrak{x}, \mathfrak{y})$ is a bilinear form on $U = V$, $W = V$, then because of its linearity properties it can be written in the form

$$b\,(\mathfrak{x}, \mathfrak{y}) = b\,(x^i\,\mathfrak{n}_i,\, y^k\,\mathfrak{n}_k) = b_{ik}\,x^i\,y^k, \tag{8.12}$$

where

$$b_{ik} = b\,(\mathfrak{n}_i, \mathfrak{n}_k). \tag{8.13}$$

Because of the same linearity properties, these components b_{ik} obey the following law of transformation under a change of basis $\bar{\mathfrak{n}}_j = a_j^{\ i}\,\mathfrak{n}_i$ in V:

$$\bar{b}_{jl} = b(\bar{\mathfrak{n}}_j, \bar{\mathfrak{n}}_l) = b(a_j^{\ i}\,\mathfrak{n}_i,\, a_l^{\ k}\,\mathfrak{n}_k) = a_j^{\ i}\,a_l^{\ k}\,b_{ik}. \tag{8.14}$$

A bilinear form $b(\mathfrak{x}, \mathfrak{y})$ of this sort is called a *doubly covariant* tensor, or a *covariant tensor of the second order*. For each subscript in (8.14), there is a transformation matrix that agrees with the corresponding transformation matrix for the components of a covariant vector in (8.11) [that is, with the second matrix in (8.11)].

Products $b(\mathfrak{x}, \mathfrak{y}) = l(\mathfrak{x})m(\mathfrak{y})$ of two linear forms on V serve as examples of covariant tensors of the second order. For the components, we have, in this case,

$$b_{ik} = b(\mathfrak{n}_i, \mathfrak{n}_k) = l(\mathfrak{n}_i)m(\mathfrak{n}_k) = l_i\, m_k,$$

showing that the tensor components here are precisely all the products of the components of the two covariant vectors $l(\mathfrak{x})$ and $m(\mathfrak{y})$.

(b) Similarly, we obtain for the components of a bilinear form $b(\mathfrak{x}, l)$ on $U = V$, $W = V_1$ that

$$\bar{b}_j{}^l = b(\bar{\mathfrak{n}}_j, \bar{\mathfrak{n}}^l) = b(a_j{}^i\, \mathfrak{n}_i, \alpha_k{}^l\, \mathfrak{n}^k) = a_j{}^i\, \alpha_k{}^l\, b_i{}^k. \tag{8.15}$$

Bilinear forms of this sort are called *simply covariant, simply contravariant* tensors, or *mixed tensors of the second order.*

(c) Finally, in case $U = W = V_1$, we obtain

$$\bar{b}^{jl} = b(\bar{\mathfrak{n}}^j, \bar{\mathfrak{n}}^l) = b(\alpha_i{}^j\, \mathfrak{n}^i, \alpha_k{}^l\, \mathfrak{n}^k) = \alpha_i{}^j\, \alpha_k{}^l\, b^{ik} \tag{8.16}$$

and we call a bilinear form of this type a *doubly contravariant* tensor, or *contravariant tensor of the second order.* To obtain an example analogous to the one under (a) above, let $\mathfrak{x}$ and $\mathfrak{y}$ be two contravariant vectors with components x^i and y^k, respectively, and set $b^{ik} = x^i y^k$; then

$$b(l, m) = b^{ik}\, l_i\, m_k$$

is indeed a contravariant tensor of the second order. The proof of this consists first in the verification of the required linearity properties, which is immediate, and then in showing that the bilinear form is independent of the basis. To show this, we note that, by (8.16) and (8.4), we have

$$\bar{b}^{jr}\, \bar{l}_j\, \bar{m}_r = b^{ik}\, \alpha_i{}^j\, \alpha_k{}^r\, a_j{}^s\, a_r{}^t\, l_s\, m_t = b^{ik}\, \delta_i^s\, \delta_k^t\, l_s\, m_t = b^{ik}\, l_i\, m_k,$$

which shows that the value of $b(l, m)$ does not depend on the choice of a particular basis.

We define the *outer product* $\mathfrak{x} \wedge \mathfrak{y}$ of two contravariant vectors[1] as the contravariant tensor of the second order whose components are

$$s^{ik} = x^i\, y^k - x^k\, y^i. \tag{8.17}$$

[1] This tensor is not to be confused with the vector product of two vectors of three-dimensional Euclidean space, which is written $\mathfrak{x} \times \mathfrak{y}$. However, note the connection pointed out in Example 1.

We verify that this is indeed a contravariant tensor of the second order by looking at the law of transformation of its components:

$$\bar{\mathfrak{s}}^{jr} = \bar{x}^j \bar{y}^r - \bar{x}^r \bar{y}^j = \alpha_i^{\ j} \alpha_k^{\ r} (x^i y^k - x^k y^i).$$

Example 1. *The vector product of two vectors of a three-dimensional vector space.* If $n = 3$, the tensor (8.17) has only three essential components (in the sense that the remaining six are three zeros and the three negatives of the first three). We denote them by

$$z_1 = x^2 y^3 - x^3 y^2, \qquad z_2 = x^3 y^1 - x^1 y^3, \qquad z_3 = x^1 y^2 - x^2 y^1$$

and correspondingly in a new coordinate system. Let us investigate how these components are transformed under a change of basis:

$$\begin{aligned}\bar{z}_1 &= \bar{x}^2 \bar{y}^3 - \bar{x}^3 \bar{y}^2 = \overset{2}{\alpha}_j \overset{3}{\alpha}_k (x^j y^k - x^k y^j) \\ &= z_1(\overset{2}{\alpha}_2 \overset{3}{\alpha}_3 - \overset{2}{\alpha}_3 \overset{3}{\alpha}_2) + z_2(\overset{2}{\alpha}_3 \overset{3}{\alpha}_1 - \overset{2}{\alpha}_1 \overset{3}{\alpha}_3) + z_3(\overset{2}{\alpha}_1 \overset{3}{\alpha}_2 - \overset{2}{\alpha}_2 \overset{3}{\alpha}_1).\end{aligned}$$

The parenthesized expressions in the second line are the two-rowed minors of the matrix $(\alpha_j^{\ k})$ and can therefore be expressed in terms of the elements of the inverse matrix $(a_i^{\ k})$, as follows:

$$\bar{z}_1 = \frac{1}{\det(a_i^{\ k})} a_1^{\ j} z_j.$$

The corresponding calculations for $\bar{z}_2$ and $\bar{z}_3$, left to the reader as an exercise, yield

$$\bar{z}_r = \frac{1}{\det(a_i^{\ k})} a_r^{\ j} z_j,$$

and we obtain the following result:

Assume that in three-dimensional space we restrict ourselves only to such changes of basis as have determinant $+1$, *that is, to transformations that preserve volume and orientation. Then it is possible to assign to the outer product of two contravariant vectors a uniquely determined covariant vector. In particular, in three-dimensional Euclidean space, where the difference between covariance and contravariance disappears if only orthonormal basis transformations are admitted* (see Exercise 8.2.2), *the result assumes the following form: The outer product can be represented as a vector if rotations of the coordinate system are the only allowed basis transformations, and as the negative of this vector if reflections* $[\det(a_i^{\ k}) = -1]$ *are allowed.*

This assignment of a vector to the outer product of two vectors is possible only for $n = 3$, since the number $\binom{n}{2}$ of independent components of the outer product equals the dimension n only for $n = 3$.

Example 2. *Infinitesimal Rotations.* Assume that an inner product

$$\mathfrak{x} \cdot \mathfrak{y} = \delta_{ik} x_i y_k$$

is defined in the vector space. We shall confine ourselves to orthonormal basis transformations only, so that all indices are written as subscripts. Let $b_{ik}(t)$ be a family of rotations such that $b_{ik}(0) = \delta_{ik}$ (the identity mapping) and such that the derivatives $d_{ik} = \dot{b}_{ik}(0)$ exist. Then the image vectors

$$x_i(t) = b_{ij}(t)\, x_j(0), \qquad y_k(t) = b_{kl}(t)\, y_l(0)$$

must satisfy

$$\delta_{ik}\, x_i(t)\, y_k(t) = \delta_{ik}\, x_i(0)\, y_k(0)$$

and therefore, in particular,

$$\begin{aligned} 0 &= \frac{d}{dt} (\delta_{ik}\, x_i(t)\, y_k(t)) \Big|_{t=0} \\ &= \delta_{ik}(d_{ij}\, \delta_{kl} + \delta_{ij}\, d_{kl})\, x_j(0)\, y_l(0) \\ &= (d_{lj} + d_{jl})\, x_j(0)\, y_l(0). \end{aligned}$$

Since the vectors $\mathfrak{x}(0)$, $\mathfrak{y}(0)$ can be chosen at will, it follows that the matrix (d_{jl}) of the infinitesimal rotation is skew symmetric:

$$d_{jl} = -\, d_{lj}.$$

Exercise 8.3.1. In Example 2, let $n = 3$ and set $\omega_1 = d_{23}$, $\omega_2 = d_{31}$, $\omega_3 = d_{12}$. Show that, under rotations of the coordinate system, the ω_i are transformed like a vector and that $\omega_i\, \mathfrak{n}_i$ can be interpreted as the angular velocity vector.

8.4 Symmetric Bilinear Forms and Inner Products

On account of their special importance, we shall study in greater detail those symmetric covariant tensors of the second order

$$b\,(\mathfrak{x}, \mathfrak{y}) = b\,(\mathfrak{y}, \mathfrak{x})$$

whose associated quadratic form is positive-definite, i.e., for which $b(\mathfrak{x}, \mathfrak{x}) > 0$ for all $\mathfrak{x} \neq 0$. Instead of $b(\mathfrak{x}, \mathfrak{y})$, we use the shorter notation $\mathfrak{x}\mathfrak{y}$ for such bilinear forms and we refer to them as *inner products*. Their associated quadratic form $\mathfrak{x}\mathfrak{x}$ is called the square of the *length* of $\mathfrak{x}$. In a given basis $\mathfrak{n}_i$ of V, the components g_{ik} of the tensor are $g_{ik} = \mathfrak{n}_i\, \mathfrak{n}_k$, by (8.13). It is known from linear algebra that $\det(g_{ik}) \neq 0$ and that there exist numbers g^{ik} such that $g^{ik} = g^{ki}$ and

$$g_{ij}\, g^{jk} = \delta_i^{\ k}. \tag{8.18}$$

The g^{ik} are the elements of the inverse matrix and can be computed in the well-known manner from the $(n - 1)$-rowed minors of g_{ik}. We use them to introduce in the dual space an inner product, called the *adjoint* of the original one and defined by

$$l \cdot m = g^{ik}\, l_i\, m_k.$$

The mapping $x^i \to g_{ik} x^k$ from the vector space V to its dual space is one-to-one, by (8.18), and the inverse mapping is $m_i \to g^{ik} m_k$.

The following representation of the g^{ik} is of special interest. Let $\mathfrak{z}_{(1)}, \ldots, \mathfrak{z}_{(n)}$ be n mutually orthogonal unit vectors, i.e., let $\mathfrak{z}_{(i)} \cdot \mathfrak{z}_{(k)} = \delta_{(ik)}$. (The subscripts are parenthesized to emphasize that they are numerical labels that have nothing to do with the behavior of the labeled entities under basis transformations.) We shall prove that

$$g^{ik} = \delta^{rs} z^i_{(r)} z^k_{(s)}. \tag{8.19}$$

To this end, we note that the $\mathfrak{z}_{(i)}$ constitute a basis of the space V, so that every vector $\mathfrak{x}$ of V has, besides its original representation $\mathfrak{x} = x^i \mathfrak{n}_i$, also a representation as

$$\mathfrak{x} = \xi^i \mathfrak{z}_{(i)} = \xi^i z^k_{(i)} \mathfrak{n}_k.$$

Now we consider the quantities

$$b^i_j = \delta^{rs} z^i_{(r)} z^k_{(s)} g_{kj}$$

and note that to prove (8.19) it is sufficient to show that $b^i_j = \delta^i_j$. But

$$b^i_j x^j = \delta^{rs} z^i_{(r)} z^k_{(s)} g_{kj} \xi^l z^j_{(l)} = \delta^{rs} \delta_{sl} z^i_{(r)} \xi^l = z^i_{(l)} \xi^l = x^i,$$

which indeed shows that $b^i_j = \delta^i_j$.

Finally, we wish to recall to the reader the well-known orthogonalization process due to Erhard Schmidt. Let $\mathfrak{x}_1, \ldots, \mathfrak{x}_n$ be n linearly independent vectors. Then all of the *Gramian* determinants

$$G_{(k)} = \begin{vmatrix} \mathfrak{x}_1 \mathfrak{x}_1 & \mathfrak{x}_1 \mathfrak{x}_2 & \cdots & \mathfrak{x}_1 \mathfrak{x}_k \\ \cdots & & & \\ \mathfrak{x}_k \mathfrak{x}_1 & & \cdots & \mathfrak{x}_k \mathfrak{x}_k \end{vmatrix}$$

are positive (for $1 \leq k \leq n$), and the n vectors

$$\mathfrak{y}_k = \frac{1}{\sqrt{G_{(k)} G_{(k-1)}}} \begin{vmatrix} \mathfrak{x}_1 \mathfrak{x}_1 \cdots \mathfrak{x}_1 \mathfrak{x}_{k-1} & \mathfrak{x}_1 \\ \mathfrak{x}_2 \mathfrak{x}_1 \cdots \mathfrak{x}_2 \mathfrak{x}_{k-1} & \mathfrak{x}_2 \\ \cdots & \\ \mathfrak{x}_k \mathfrak{x}_1 \cdots \mathfrak{x}_k \mathfrak{x}_{k-1} & \mathfrak{x}_k \end{vmatrix}$$

are an orthonormal set: $\mathfrak{y}_i \mathfrak{y}_k = \delta_{ik}$. The $\mathfrak{y}_k$ are constructed, one by one, as follows: We first set $\mathfrak{y}_1 = \mathfrak{x}_1 / \sqrt{\mathfrak{x}_1 \mathfrak{x}_1}$. Next, $\mathfrak{y}_2$ is a unit vector that lies in the plane spanned by $\mathfrak{x}_1$, $\mathfrak{x}_2$ and is orthogonal to $\mathfrak{y}_1$. More generally, $\mathfrak{y}_k$ is a unit vector that lies in the linear subspace spanned by $\mathfrak{y}_1, \ldots, \mathfrak{y}_{k-1}, \mathfrak{x}_k$ and is orthogonal to all $\mathfrak{y}_j$ with $j < k$.

8.5 Tensors of Any Order

Definition. *A form (functional) t that is multilinear in r contravariant and s covariant vectors is called an r-tuply covariant, s-tuply contravariant tensor, or a mixed tensor of order*[2] $r + s$, *or even more briefly, an* (r, s)*-tensor. Relative to a given basis* $\mathfrak{n}_i$ *for the vector space V and the dual basis* $\mathfrak{n}^k$ *for the dual space, the components of such a tensor are obtained by substitution of the basis vectors into the multilinear form:*

$$t\,{}^{i_1, \ldots, i_s}_{k_1, \ldots, k_r} = t\,(\mathfrak{n}^{i_1}, \ldots, \mathfrak{n}^{i_s};\quad \mathfrak{n}_{k_1}, \ldots, \mathfrak{n}_{k_r}).$$

We define the form t to be $= 0$ *if the value of t is* $= 0$ *for all systems of its* $r + s$ *argument vectors. Clearly, a necessary and sufficient condition for this is that, relative to some basis, all components of the tensor should be* $= 0$. *(They are then* $= 0$ *relative to every basis.)*

Under a change of basis, each superscript (or subscript, resp.) is transformed like a superscript (or subscript, resp.) on a component of a contravariant vector (or covariant vector, respectively). The proof of this is similar to that in the case of tensors of the second order. Conversely, we can see in the same way as for second-order tensors that, if components are given that satisfy the appropriate transformation law under a change of basis, then these components belong to an invariant multilinear form.

For the sake of completeness, contravariant vectors are also called (0, 1)-tensors, covariant vectors are called (1, 0)-tensors, and constants (scalars) are called (0, 0)-tensors. The pair (r, s) is the *order type* of the tensor.

Tensors are subject to the following obvious rules of calculation:

1. Tensors of the same order type are *added* by adding the values of the two multilinear forms at like sets of argument vectors. Thus — using tensors of low order to illustrate, since things are no different for higher orders —

$$(t + u)\,(\mathfrak{l};\, \mathfrak{x}, \mathfrak{y}) \underset{\text{def}}{=} t\,(\mathfrak{l};\, \mathfrak{x}, \mathfrak{y}) + u\,(\mathfrak{l};\, \mathfrak{x}, \mathfrak{y}).$$

In terms of components, we must add corresponding components to obtain the components of the sum tensor.

2. Tensors of any order types can be multiplied and the product of an (r, s)-tensor by an (r', s')-tensor is an $(r + r', s + s')$-tensor:

$$(t \cdot v)\,(\mathfrak{l}, \mathfrak{m};\, \mathfrak{x}, \mathfrak{y}, \mathfrak{z}) \underset{\text{def}}{=} t\,(\mathfrak{l};\, \mathfrak{x}, \mathfrak{y})\, v\,(\mathfrak{m};\, \mathfrak{z}).$$

---

[2] Unless $r = 0$ or $s = 0$, in which cases one speaks of a contravariant (or covariant, resp.) tensor of order s (or r, resp.).

The components of the product $w = t \cdot v$ are all the possible product of components of t by components of v:

$$w^{ij}_{klm} = t^{i}_{kl}\, v^{j}_{m}.$$

The multiplication of tensors by scalars is contained in this definition as a special case.

Multiplication and addition are commutative: $t \cdot v = v \cdot t$; $t + u = u + t$, and the distributive law holds: $v(u + t) = v \cdot u + v \cdot t$; likewise, the associative laws, e.g., $t(uv) = (tu)v$. All of these follow from the validity of the corresponding laws in the domain of real numbers. It follows that the tensors of a fixed order type constitute a real vector space.

3. There is a less obvious operation, not binary, but unary, which, applied to an (r, s)-tensor, yields an $(r - 1, s - 1)$-tensor and which is called *contraction*. It is easiest to define this operation in terms of the tensor components relative to some given basis and to show afterward its independence of the basis that was used. Let t be a mixed tensor ($r \geqq 1$, $s \geqq 1$), say a (2, 1)-tensor; its components are then transformed in accordance with

$$\bar{t}^{i}_{jl} = a_j{}^k\, a_l{}^m\, t^{p}_{km}\, \alpha_p{}^i.$$

Now we define the tensor s obtained from t by contraction with respect to p and m as the tensor whose components, in the given base, are

$$s_k = t^{m}_{km},$$

where summation over m is indicated. Noting that $a_l{}^m\, \alpha_m{}^i = \delta^i_l$, we obtain for the tensor s the transformation law

$$\bar{s}_j = \bar{t}^{i}_{ji} = a_j{}^k\, a_i{}^m\, t^{p}_{km}\, \alpha_p{}^i = a_j{}^k\, \delta^m_p\, t^{p}_{km} = a_j{}^k\, s_k.$$

Thus the definition, in terms of tensor components, is as follows: Contraction of a tensor with respect to a superscript p and a subscript m means that the components with $p = m$ are to be summed, while all remaining indices remain fixed.

4. A particularly important case of contraction is involved in the *inner multiplication* of two tensors t and u. Their inner product with respect to a superscript of the one and a subscript of the other is defined as the tensor obtained from their product tensor $t \cdot u$ by the appropriate contraction, thus, for instance:

$$t^{i}_{kj}\, v^{j}_{m}.$$

By what was proved under 3 above, such inner products are themselves tensors. An important example is furnished by Eq. (8.18).

The significance of these algebraic operations on tensors lies in the fact that, with their help, we are able to produce new tensors from given tensors. Contraction, in particular, can lead to invariants, since it decreases the order of a tensor.

It is common usage to apply the term "tensor," not only to the multilinear form itself, but also to the totality of its components relative to a particular basis and thus to speak of, say, "the tensor g_{ik}." Even though it is not entirely unexceptionable, we shall often follow this usage because of its convenience.

One of the frequent objections against this usage is that formulas involving tensors are then not written invariantly, but in a way that is dependent on the choice of a particular basis. R. and F. Nevanlinna, in their book "Absolute Analysis" ("Grundlehren," Vol. 102, Springer, Berlin-Göttingen-Heidelberg, 1959), have therefore proposed an "index-free" development of the tensor calculus, avoiding the use of particular bases altogether. It turns out, however, that this makes the written description of the tensor calculus lose much of its visual convenience. Thus, all of the information that the number and position of the indices conveys has to be separately given in the text if an index-free notation is adopted. Moreover it is not easy to apply the index-free formulas to concrete problems, say, in the theory of surfaces. Indeed, in Chapter V entitled Differential Geometry of the above-mentioned book by the Nevanlinnas, there are no investigations of particular surfaces at all. It is possible, incidentally, to interpret the notation with indices in a way that is independent of any basis and, thus, to combine the practical advantages of indicial notation with the conceptual advantages of independence from bases [see D. Laugwitz, *Math. Z.* **61**, 100 and 134 (1954)].

Exercise 8.5.1. Define inner multiplication of tensors in a basis-free way. [*Hint.* Consider, say, the example $t_{kj}^{i} v_m^{j}$. The bilinear form $v = v(\,.\,; \mathfrak{y})$ is a contravariant vector whenever its second argument is held fixed, since it is then a linear form in a covariant vector. We substitute this contravariant vector into the last argument place of the trilinear form t and we obtain

$$t(l; \mathfrak{x}, v(\,.\,; \mathfrak{y})),$$

and this is an invariant trilinear form whose components are $t_{kj}^{i} v_m^{j}$.

§ 9. Tensor Analysis

9.1 Introduction

In the theory of surfaces, we already had occasion to deal with tensor fields, such as g_{ik} or L_{ik}, which assign to every point of a manifold a tensor of a fixed order type. In the framework of differential geometry it is almost self-evident that one needs derivatives of such tensor fields. The problem arises of defining the derivative of a tensor field in such a way that this derivative is itself a tensor field because otherwise one could not be sure that the derivative would have a meaning that is independent of the coordinate system.

Let P_0 be a fixed point and assume that in a neighborhood of P_0, a coordinate system (x^i) is given, with P_0 itself having coordinates $\underset{0}{x}^i$. Then, in the tangent space at P_0, a particular basis is thereby made distinguished: The distinguished basis vectors $\mathfrak{n}_i$ are the tangent vectors of the curves $x^k(t) = \underset{0}{x}^k$ for $k \neq i$, $x^i(t) = t + \underset{0}{x}^i$. A second coordinate system $(\bar{x}^i)$ similarly determines its own distinguished basis. For the change of basis from the first to the second, let us compute the transformation equations (8.2). To this end let $x^i(t)$ be any curve and let $\bar{x}^k(t)$ be the representation of this curve in the new coordinate system. Then $x^i(t) = x^i(\bar{x}^k(t))$ and thus, by the chain rule,

$$\dot{x}^i(t) = \frac{\partial x^i}{\partial \bar{x}^k} \dot{\bar{x}}^k(t). \tag{9.1a}$$

Since the tangent vector at P_0 is, by its definition, independent of the basis, we must have

$$\mathfrak{n}_i \dot{x}^i = \bar{\mathfrak{n}}_k \dot{\bar{x}}^k,$$

and therefore, substituting from (9.1a) and (8.2),

$$\mathfrak{n}_i \frac{\partial x^i}{\partial \bar{x}^k} \dot{\bar{x}}^k = a_k{}^i \mathfrak{n}_i \dot{\bar{x}}^k,$$

and since this holds for arbitrary $\dot{\bar{x}}^k$, it follows that

$$a_k{}^i = \frac{\partial x^i}{\partial \bar{x}^k}. \tag{9.1b}$$

In the same way, we find, for the inverse transformation (8.3), that

$$\alpha_j{}^k = \frac{\partial \bar{x}^k}{\partial x^j}. \tag{9.1c}$$

Formulas (9.1b) and (9.1c) are basic for the application of tensor algebra to the tangent space of a differentiable manifold at a point P_0.

Now let us begin with a tensor field of the simplest type, namely, a scalar field $\varphi(x^i) = \bar{\varphi}(\bar{x}^i)$. By the chain rule,

$$\varphi_{,j} = \bar{\varphi}_{,k} \frac{\partial \bar{x}^k}{\partial x^j},$$

where subscripts preceded by a comma denote partial differentiation. This shows that *the gradient field* $\varphi_{,j} = \partial\varphi/\partial x^j$ *of a scalar field* $\varphi(x^j)$ *is a covariant vector field.* In this case, therefore, our problem is already solved: The usual partial differentiation operation furnishes a tensor field and is, therefore, a usable covariant derivative. Unfortunately, this same simple procedure is not successful for tensors of higher-order types. Even for a contravariant vector field $\xi^i(x^j)$, differentiation with respect to $\bar{x}^j$ of the transformation formula

$$\bar{\xi}^i = \xi^k \frac{\partial \bar{x}^i}{\partial x^k}$$

yields

$$\bar{\xi}^i_{,j} = \xi^k_{,r} \frac{\partial \bar{x}^i}{\partial x^k} \frac{\partial x^r}{\partial \bar{x}^j} + \xi^k \frac{\partial^2 \bar{x}^i}{\partial x^k\, \partial x^r} \frac{\partial x^r}{\partial \bar{x}^j}\,. \tag{9.2}$$

If the second term on the right-hand side were not present — which is the case if and only if the coordinate transformations are linear — then $\xi^k_{,r}$ would indeed obey the transformation law of a (1, 1)-tensor. As it is, however, the usual partial differentiation operation applied to a contravariant vector field has no invariant, "tensorial" meaning under general coordinate transformations. We shall have to modify the operation in such a way as to preserve some important properties of the usual partial differentiation, while at the same time insuring that the modified operation, so-called covariant differentiation, obeys the transformation law of a tensor.

9.2 Covariant Differentiation of a Contravariant Vector Field

In view of the above, the idea suggests itself to try to compensate by means of a correction term for $\xi^i_{,j}$ for the deviation noted in (9.2) from the desired tensorial law of transformation. To insure for the corrected, *covariant derivative* $\xi^i_{;j}$ that it satisfy $(\xi^i + \eta^i)_{;j} = \xi^i_{;j} + \eta^i_{;j}$, we shall try a correction term that is linear in ξ^i:

$$\xi^i_{;j} = \xi^i_{,j} + \Lambda^i_{jk}\,(x)\,\xi^k. \tag{9.3}$$

The functions Λ^i_{jk} are quite arbitrary to begin with, and we must find out to what requirements they are to be subjected so that $\xi^i_{;j}$ will be a tensor for any choice of the vector field ξ^i. Setting

$$\bar{\xi}^i_{;j} = \bar{\xi}^i_{,j} + \bar{\Lambda}^i_{jk} \bar{\xi}^k,$$

we can state the requirement of a tensorial transformation law in the form

$$\bar{\xi}^i_{;j} = \frac{\partial \bar{x}^i}{\partial x^k} \frac{\partial x^r}{\partial \bar{x}^j} \xi^k_{;r}$$

and we find, after substituting from (9.3) and (9.2), that we must have

$$\bar{\Lambda}^i_{js} \bar{\xi}^s = \Lambda^k_{rl} \xi^l \frac{\partial \bar{x}^i}{\partial x^k} \frac{\partial x^r}{\partial \bar{x}^j} - \xi^l \frac{\partial^2 \bar{x}^i}{\partial x^l \partial x^r} \frac{\partial x^r}{\partial \bar{x}^j},$$

or, since this has to hold for all ξ^l,

$$\bar{\Lambda}^i_{js} \frac{\partial \bar{x}^s}{\partial x^l} = \Lambda^k_{rl} \frac{\partial \bar{x}^i}{\partial x^k} \frac{\partial x^r}{\partial \bar{x}^j} - \frac{\partial^2 \bar{x}^i}{\partial x^l \partial x^r} \frac{\partial x^r}{\partial \bar{x}^j}. \tag{9.4}$$

This shows that the quantities Λ^i_{jk} are not the components of a tensor, on account of the last term in (9.4) which contains the second derivatives of the coordinate transformation. But this term is there as part of the requirement that the partial derivatives $\xi^i_{,j}$ together with the correction term should constitute a tensor $\xi^i_{;j}$.

By going through the calculations in the reverse order, we can verify that every system Λ^i_{jk} satisfying the transformation law (9.4) leads to a tensor (9.3), a tensor which can therefore be called a covariant derivative of the vector field ξ^i. To be sure, this differentiation process is not uniquely determined since the Λ^i_{jk} are still largely arbitrary; in fact, we can choose them entirely at will in a given coordinate system and then obtain $\bar{\Lambda}^i_{js}$ for all other coordinate systems from (9.4). This arbitrariness in the definition of a covariant derivative is not a disadvantage at all. We shall see that, in concrete cases, it is possible to select, from the variety of available covariant differentiations, one which is particularly well suited to the problem at hand and which is then usually uniquely determined thereby.

The example for a covariant differentiation which was historically the earliest, and which, without doubt, still remains the most important, is one which we have already encountered in the theory of surfaces: The Christoffel symbols Γ^i_{jk} obey precisely the transformation law (9.4) (see §4.2). Thus there is, for every vector field $\mathfrak{y} = \xi^i \mathfrak{x}_i$ defined on a surface, a covariant derivative

$$\xi^i_{;j} = \xi^i_{,j} + \Gamma^i_{jk} \xi^k.$$

9.3 Geometric Interpretation of Covariant Differentiation: Parallel Displacement

In the special case of the theory of surfaces, there is associated with the Γ^i_{jk} the *parallel displacement of Levi-Cività* which also yields a geometric interpretation of covariant differentiation: If $x^i(t)$ is a curve on a surface and if covariant differentiation along this curve is defined by

$$\frac{D\xi^i}{dt} = \xi^i_{;j}\,\dot{x}^j = \frac{d\xi^i}{dt} + \Lambda^i_{jk}\,\xi^k\,\dot{x}^j,$$

then, taking $\Lambda^i_{jk} = \Gamma^i_{jk}$, the condition for $\xi^i(t)$ to be a vector field of parallel vectors in the sense of Levi-Cività, is that

$$D\xi^i/dt = 0.$$

This geometric interpretation in the case of the theory of surfaces suggests the following definition for the case of a general covariant differentiation process also: A vector field $\xi^i(x)$ is said to be parallel along a curve $x^i(t)$ if

$$\frac{D\xi^i}{dt} \underset{\text{def}}{=} \xi^i_{;j}\,\dot{x}^j = 0. \tag{9.5}$$

With this definition we can accomplish, just as in the theory of surfaces, a parallel displacement along a curve $x^i(t)$, with $x^i(t_0) = \underset{0}{x}^i$, of the entire tangent space T_0 at a point $(\underset{0}{x}^i)$. This is done by assigning to each vector ξ^i of T_0 the solution of the system of differential equations (9.5) that satisfies the initial condition $\xi^i(t_0) = \underset{0}{\xi}^i$. This parallel[2] displacement, defined in terms of the Λ^i_{jk}, induces a linear mapping

$$\xi^i(t) = A^i_k(t)\,\xi^k(t_0)$$

from the tangent space T_0 into the tangent space at the point $(x(t))$, since the system (9.5) is linear in ξ. Note that this notion of parallel displacement depends on the Λ^i_{jk} as well as on the curve $x^i(t)$.

---

[2] The term parallel is here, as it was in surface theory, used as short for "obtainable from each other by parallel displacement." Accordingly, vectors of ordinary affine space are parallel if and only if they are componentwise the same. Note the distinction between this notion of parallelism and the one used synonymously with "having the same direction."

We shall say that an assignment of functions Λ^i_{jk} furnishes the manifold with a *linear connection* and we call the Λ^i_{jk} the components of this linear connection. Without such an assignment, the tangent spaces at different points of the manifold are unrelated to each other, so to say, or without any "connection" to each other, while with such an assignment they can be connected by linear mappings.

9.4 *Affine Connections*

The Γ^i_{jk} of the theory of surfaces have the property of being symmetric in their two subscripts: $\Gamma^i_{jk} = \Gamma^i_{kj}$. This property of a linear connection is independent of the coordinate system, as can be seen from (9.4), and it has a significant geometric interpretation.

Theorem 9.4.1. *A linear connection is symmetric, in the sense that $\Lambda^i_{jk} = \Lambda^i_{kj}$, if and only if there is for each point P_0 a coordinate system in which all the components of the linear connection have the value 0 at P_0. Such a coordinate system is then said to be "geodesic at P_0".*

Proof. If $\Lambda^i_{jk}(\underset{0}{x}^r) = 0$ in a coordinate system x, then it follows from the transformation formulas (9.4) that $\bar{\Lambda}^i_{jk}(\underset{0}{\bar{x}}^r) = \bar{\Lambda}^i_{kj}(\underset{0}{\bar{x}}^r)$. Conversely, assume that $\bar{\Lambda}^i_{jk}(\underset{0}{\bar{x}}^r)$ is symmetric. We shall find a coordinate transformation $x^i = x^i(\bar{x}^r)$ such that $\underset{0}{x}^r = 0$ and $\Lambda^i_{jk}(0) = 0$, namely,

$$x^i = (\bar{x}^i - \underset{0}{\bar{x}}^i) + \frac{1}{2}\bar{\Lambda}^i_{jk}(\underset{0}{\bar{x}}^r)(\bar{x}^j - \underset{0}{\bar{x}}^j)(\bar{x}^k - \underset{0}{\bar{x}}^k).$$

We see, using the symmetry of the $\bar{\Lambda}^i_{jk}$, that

$$\frac{\partial x^i}{\partial \bar{x}^k} = \delta^i_k + \bar{\Lambda}^i_{jk}(\underset{0}{\bar{x}})(\bar{x}^j - \underset{0}{\bar{x}}^j),$$

which shows that the coordinate transformation is even analytic and that its functional determinant is not zero in some neighborhood of $\bar{x}^i = \underset{0}{\bar{x}}^i$, being equal to 1 at that point and being continuous. Hence, the coordinate transformation is admissible. It follows further that

$$\frac{\partial^2 x^i}{\partial \bar{x}^j\, \partial \bar{x}^k} = \bar{\Lambda}^i_{jk}(\underset{0}{\bar{x}}),$$

and substitution of this into the transformation formulas (9.4) completes the proof of the theorem.

Thus, if a linear connection is symmetric, then there is for each point a coordinate system such that the covariant derivatives at the point are simply the partial derivatives at the point. A manifold with a symmetric linear connection is said to be *affinely connected* and the symmetric linear connection itself is called an *affine connection*, because ordinary affine space does have this property. (In ordinary affine space we even have $\Lambda^i_{jk} = 0$ simultaneously for all points if the coordinate system is rectangular Cartesian, since, in this case, a parallel vector field satisfies $d\xi^l/dt = 0$.) In the sequel we shall be able to confine ourselves mainly to the consideration of affine connections, which, as mentioned above, include the connection given by Γ^i_{jk} that is basic for the theory of surfaces.

Exercise 9.4.2. In dealing with real vector spaces, one can define the usual derivative of a vector field ξ^i along a curve $x^i(t)$ by

$$\frac{d\xi^i}{dt} = \lim_{\Delta t \to 0} \frac{\xi^i(x(t + \Delta t)) - \xi^i(x(t))}{\Delta t}.$$

In a differentiable manifold, it does not, offhand, make sense to speak of subtracting vectors that are defined (see §7.3) at different points $x(t + \Delta t)$ and $x(t)$. However, show the following: If $\tilde{\xi}^i(x(t + \Delta t))$ is the vector that is obtained from $\xi^i(x(t))$ by parallel displacement along the curve $x(t)$, then

$$\frac{D\xi^i}{dt} = \lim_{\Delta t \to 0} \frac{\xi^i(x(t + \Delta t)) - \tilde{\xi}^i(x(t + \Delta t))}{\Delta t}.$$

(*Hint.* By the law of the mean of the differential calculus, we have

$$\xi^i(t + \Delta t) = \xi^i(t) + \Delta t \frac{d\xi^i(t + \vartheta_i \Delta t)}{dt}$$

and

$$\tilde{\xi}^i(t + \Delta t) = \xi^i(t) + \Delta t \frac{d\tilde{\xi}^i(t + \delta_i \Delta t)}{dt}$$

with $0 < \vartheta_i, \delta_i < 1$. Note that $d\tilde{\xi}^i/dt = -\Lambda^i_{jk} \tilde{\xi}^k \dot{x}^j$.)

9.5 Covariant Differentiation of Tensors of Any Order

So far, we have discussed covariant differentiation only for scalar fields and for contravariant vector fields:

$$\varphi_{;j} = \varphi_{,j} \tag{A}$$

$$\xi^i_{;j} = \xi^i_{,j} + \Lambda^i_{jk} \xi^k, \tag{B}$$

provided that a linear connection Λ^i_{jk} is given. By adding a single, very plausible requirement (C), we shall obtain from (A) and (B) formulas defining covariant differentiation of arbitrary tensor fields:

For multilinear forms, covariant differentiation should obey the product rule. (C)

[For (1, 1)-tensors, say:

$$(u_i t^i_k \xi^k)_{;j} = u_{i;j} t^i_k \xi^k + u_i t^i_{k;j} \xi^k + u_i t^i_k \xi^k_{;j} \qquad \text{etc.}]$$

To begin with, let us obtain from these postulates the covariant derivative of a covariant vector field u_i. The contravariant vector field ξ^i involved in the following calculation is arbitrary and the letters under signs of equality refer to the postulate from which the equality follows:

$$u_{i,j}\xi^i + u_i \xi^i_{,j} = (u_i\xi^i)_{,j} \underset{\text{A}}{=} (u_i \xi^i)_{;j} \underset{\text{C}}{=} u_{i;j}\xi^i + u_i \xi^i_{;j} \underset{\text{B}}{=} u_{i;j}\xi^i + u_i \xi^i_{,j} + \Lambda^i_{jk}\xi^k u_i.$$

Comparison of the first with the last expression yields

$$u_{k,j}\xi^k = u_{k;j}\xi^k + u_i \Lambda^i_{jk}\xi^k,$$

and, this being valid for all ξ^k, it follows that

$$u_{k;j} = u_{k,j} - \Lambda^i_{jk} u_i. \qquad \text{(D)}$$

This covariant derivative of u_k is a covariant tensor of order 2, as can be seen from the above equality $\underset{\text{C}}{=}$ whose left- and right-hand sides are covariant vectors.

For (1, 1)-tensor fields, we have similarly, with arbitrary u_i, ξ^k:

$$\begin{aligned} u_{i,j} t^i_k \xi^k + u_i t^i_{k,j} \xi^k + u_i t^i_k \xi^k_{,j} &= (u_i t^i_k \xi^k)_{,j} \\ &\underset{\text{A}}{=} (u_i t^i_k \xi^k)_{;j} \\ &\underset{\text{C}}{=} u_{i;j} t^i_k \xi^k + u_i t^i_{k;j} \xi^k + u_i t^i_k \xi^k_{;j} \\ &\underset{\text{B, D}}{=} u_{i,j} t^i_k \xi^k - \Lambda^r_{ji} u_r t^i_k \xi^k + u_i t^i_{k;j} \xi^k + u_i t^i_k \xi^k_{,j} + u_i t^i_k \Lambda^k_{js} \xi^s, \end{aligned}$$

so that, since the u_i and ξ^k are arbitrary,

$$t^i_{k;j} = t^i_{k,j} + \Lambda^i_{jr} t^r_k - \Lambda^s_{jk} t^i_s.$$

The tensor property can now be proved just as before. Since all superscripts behave alike, and all subscripts likewise, it follows more generally that

$$t^{i_1,\ldots,i_m}_{k_1,\ldots,k_n;j} = t^{i_1,\ldots,i_m}_{k_1,\ldots,k_n,j} + \sum_{\mu=1}^{m} \Lambda^{i_\mu}_{jr}\, t^{i_1,\ldots,r,\ldots,i_m}_{k_1,\ldots,k_n} - \sum_{\nu=1}^{n} \Lambda^{s}_{jk_\nu}\, t^{i_1,\ldots,i_m}_{k_1,\ldots,s,\ldots,k_n}. \tag{E}$$

Here the superscript r occurring in the first sum occupies the μth place, the subscript s occurring in the second sum occupies the νth place. (E) represents the formula for the covariant derivative of a tensor field of any type. We see also that, once a linear connection is assigned, this covariant derivative is uniquely determined by (A)—(C), and that the covariant derivatives agree with the partial derivatives in case $\Lambda^i_{jk} = 0$.

Covariant differentiation of an (r, s)-tensor yields an $(r + 1, s)$-tensor, i.e., increases by 1 the order of covariance.

Exercise 9.5.1. *Show* that a tensor need not have a vanishing covariant derivative even if the tensor has constant components in some coordinate system. However, $\delta^i_{k;j} = 0$ holds.

Exercise 9.5.2. *Show* that the product rule holds for any product of tensors. In particular, investigate $t^i_r s^r_k$ and $t^i_r s^r_i$.

9.6 *Second Covariant Derivatives and the Curvature Tensor of an Affine Connection*

By its definition, covariant differentiation of multilinear forms obeys the product rule. But not all of the rules that apply to partial derivatives carry over to covariant derivatives. In particular, we shall show that the second covariant derivative is, in general, dependent on the order in which the two differentiations are carried out, even if the second partial derivatives are continuous. We confine ourselves here to the case of affine connections, $\Lambda^i_{jk} = \Lambda^i_{kj}$.

Theorem 9.6.1. *Let ξ^i be a vector field of class C^2 on a space with affine connection $(\Lambda^i_{jk} = \Lambda^i_{kj})$. Then*

$$\xi^i_{;j;k} - \xi^i_{;k;j} = R^i{}_{hjk}\, \xi^h, \tag{9.6}$$

where

$$R^i{}_{hjk} \underset{\text{def}}{=} \Lambda^i_{hj,k} - \Lambda^i_{hk,j} + \Lambda^r_{hj}\Lambda^i_{rk} - \Lambda^r_{hk}\Lambda^i_{rj} \tag{9.7}$$

is a tensor which is called the curvature tensor *of the affine connection.*

Proof. If we apply the differentiation formula (E) of §9.5 to $t^i_j = \xi^i_{;j}$, we obtain

$$\xi^i_{;j;k} = \xi^i_{;j,k} + \Lambda^i_{kr}\,\xi^r_{;j} - \Lambda^r_{kj}\,\xi^i_{;r}$$
$$= \xi^i_{,j,k} + \Lambda^i_{jr,k}\,\xi^r + \Lambda^i_{jr}\,\xi^r_{,k} + \Lambda^i_{kr}\,\xi^r_{,j} + \Lambda^i_{kr}\Lambda^r_{js}\,\xi^s - \Lambda^r_{kj}\,\xi^i_{,r} - \Lambda^r_{kj}\Lambda^i_{rs}\,\xi^s.$$

If we subtract from this the expression that results from interchanging the subscripts j and k, and make use of the symmetry of the connection, we obtain (9.6). To verify that R^i_{hjk} is indeed a tensor, we note that the left-hand side of (9.6) is a tensor; therefore, if we successively form its inner product with arbitrary vectors u_i, η^j, ζ^k, it becomes an invariant and the right-hand side is then an invariant multilinear form in u, ξ, η, ζ.

The name "curvature tensor" is justified in terms of the theory of surfaces, where the curvature tensor of the Γ^i_{jk} is intimately related to the Gaussian curvature K; for we have, by §5.5:

$$K \cdot g = g_{2j}\,R^j{}_{112}.$$

This example from the theory of surfaces shows at the same time that the curvature tensor does not vanish, in general, as $K = 0$ holds only for developable surfaces.

Exercise 9.6.2. *Compute* $\xi^i_{;j;k} - \xi^i_{;k;j}$ for general (i.e., not necessarily symmetric) Λ^i_{jk}. *Show* that the quantities $\Lambda^i_{jk} - \Lambda^i_{kj} = C^i_{jk}$ that occur in this computation are the components of a tensor (the *torsion tensor* of the linear connection).

Exercise 9.6.3. *Show* that

$$u_{i;j;k} - u_{i;k;j} = -\,R^l{}_{ijk}\,u_l$$

holds for affine connections.

9.7 Symmetry Properties of the Curvature Tensor

The components of the curvature tensor are not all independent. Even from the definition (9.7), it follows that

$$R^i{}_{hjk} + R^i{}_{hkj} = 0. \tag{9.8}$$

The most convenient method for obtaining further identities is to use a coordinate system that is geodesic at the point in question, so that all Λ^i_{jk} vanish at this point. By §9.4, such a coordinate system always exists in the (present) case of an affine connection. Using it, we have, by (9.7), that

$$R^i{}_{hjk} + R^i{}_{jkh} + R^i{}_{khj} = 0 \tag{9.9}$$

and since this is a tensor equation it must hold in every coordinate system. Once more using coordinates geodesic at the point, so that the covariant derivative at the point agrees with the partial derivatives, we obtain

$$R^i{}_{hjk;l} + R^i{}_{hkl;j} + R^i{}_{hlj;k} = 0. \tag{9.10}$$

This is known as Bianchi's identity. All of these relations will be used frequently in the sequel.

Exercise 9.7.1. *Prove*: For dimension $n = 2$, relations (9.9) are a consequence of (9.8).

§ 10. The Geometry of a Space with Affine Connection

Up to this point we have developed tensor analysis primarily from a formal point of view: We were motivated by the goal of constructing a generalization of the differential calculus that would be applicable to differentiable manifolds, that would preserve important properties of partial differentiation, and that would, moreover, map tensor fields upon tensor fields by means of "covariant" differentiation. It turned out that such a covariant differentiation is not determined *a priori* when the manifold is given, but that it requires the imposition upon the manifold of an additional structure, a linear connection, whose geometric counterpart is the possibility of an infinitesimal parallel displacement of vectors. In the present section we shall study the geometric properties of such a parallel displacement in greater detail, confining ourselves for this purpose to the case of a symmetric connection. The symmetric case is the only one we shall need for later applications, and it is also the most interesting one from the geometric point of view because a space with an affine connection behaves, at least "in the infinitesimally small," just like ordinary affine space.

10.1 Parallel Displacement along Closed Paths

The infinitesimal parallel displacement associated with a linear connection can be used, in particular, to displace the tangent space $T(P)$ at a point P along closed paths W that begin and end at P. This will induce a linear homogeneous mapping A_W of the vector space $T(P)$ onto itself. We have already met with examples of this state of affairs in the theory of surfaces (§4.7, in particular, Exercise 4.7.4).

The path W will here be assumed to be continuous (i.e., to be the continuous image of a real line segment) and to be pieced together from finitely many differentiable pieces. In particular, self-intersections are allowed for W, and the point P itself does not have to be a simple point of W.

The linear mapping A_W is obtained as follows. Let $\underset{0}{\xi}^i$ be any vector of $T(P)$. We obtain a vector field parallel to $\underset{0}{\xi}^i$ along the first of the smooth pieces of W by solving the differential equations

$$\frac{D\xi^i}{dt} = 0$$

along this piece of the path, subject to the initial conditions $\xi^i(0) = \underset{0}{\xi}^i$. We proceed in the same way for the second (adjoining) smooth piece of the path, except that this time we use for our initial conditions the field vector at the end point of the first smooth piece. Continuing in this way, we obtain a vector $\overset{0}{\xi}{}^i$ at the end point of the path W. If this end point is the same as the initial point P, then $\underset{0}{\xi}^i \to \overset{0}{\xi}{}^i$ is a mapping of $T(P)$ onto itself which is linear because all of the differential equations to be integrated were linear (see Fig. 28):

$$\overset{0}{\xi}{}^i = A^i_k\,(W)\,\underset{0}{\xi}^k.$$

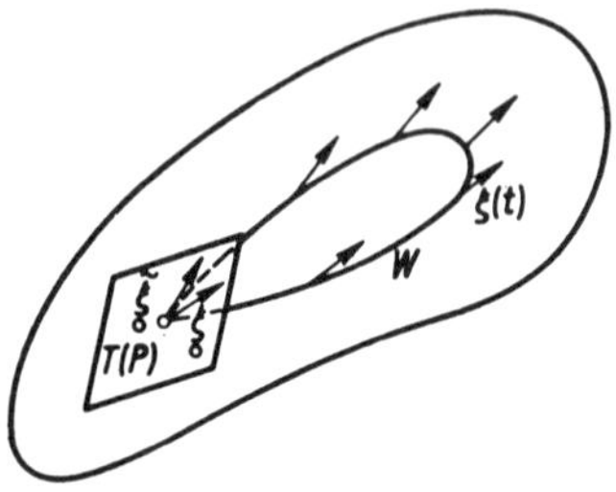

FIG. 28. Parallel displacement.

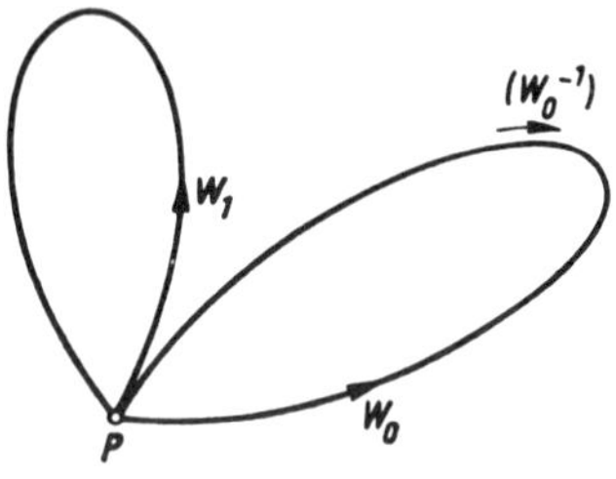

FIG. 29. The group of paths.

The mapping is independent of the coordinate system since $D\xi^i/dt$ is a vector and since, therefore, all integrations led to invariant results. Hence, introducing component-free notation for the vectors of $T(P)$ and denoting the linear mapping just discussed by $\mathfrak{A}(W)$, we can write the mapping, in invariant (i.e., basis-free) notation,

$$\bar{\mathfrak{x}} = \mathfrak{A}(W)\,\mathfrak{x}.$$

We now define a "multiplication" of paths that begin and end at P, as follows: $W = W_1 W_0$ denotes the (oriented) path that is traversed if first W_0 and then W_1 is traversed. We also define W^{-1} to mean the path obtained from W by reversing its orientation (see Fig. 29). The associated linear mappings of $T(P)$ upon itself clearly satisfy

$$\mathfrak{A}(W_1 W_0) = \mathfrak{A}(W_1)\,\mathfrak{A}(W_0) \qquad \text{and} \qquad \mathfrak{A}(W^{-1}) = [\mathfrak{A}(W)]^{-1}.$$

If we let W range over all possible paths in the manifold that begin and end at P, then the totality $H(P)$ of the associated linear mappings $\mathfrak{A}(W)$ has the following properties:

1. The identity mapping $\mathfrak{E}$ (defined by $\mathfrak{E}\mathfrak{x} = \mathfrak{x}$, all $\mathfrak{x}$) belongs to $H(P)$ [for, $\mathfrak{E} = \mathfrak{A}(W)\mathfrak{A}(W^{-1}) = \mathfrak{A}(WW^{-1})$].

2. If $\mathfrak{A}$ belongs to $H(P)$, then so does $\mathfrak{A}^{-1}$ [for, if $\mathfrak{A} = \mathfrak{A}(W)$ then $\mathfrak{A}^{-1} = \mathfrak{A}(W^{-1})$].

3. If $\mathfrak{A}$ and $\mathfrak{B}$ belong to $H(P)$, then so does $\mathfrak{A}\mathfrak{B}$ [for, if $\mathfrak{A} = \mathfrak{A}(W_1)$ and $\mathfrak{B} = \mathfrak{A}(W_0)$, then $\mathfrak{A}\mathfrak{B} = \mathfrak{A}(W_1 W_0)$].

But if a set of mappings has properties 1, 2, 3, then the set is a group (a transformation group). We can therefore state the following result and give the following definition:

Let P be a point of a differentiable manifold with a linear connection. Then the linear mappings of the tangent space $T(P)$ onto itself that are induced by parallel displacements along any paths beginning and ending at P constitute a group. This group is called the holonomy group of the connection at P.

The term "holonomy group" is not a happy choice, but it is in common use.

Exercise 10.1.1. *Prove*: (a) The holonomy group at any given point of the plane consists only of $\mathfrak{E}$. (b) For any given point P of the sphere, the holonomy group of geodesic parallelism on the sphere (see Exercise 4.7.4) is the full group of rotations of $T(P)$ about the origin of $T(P)$.

Exercise 10.1.2. *Prove*: If P and Q are two points of a space with an affine connection, then the holonomy groups $H(P)$ and $H(Q)$ are not essentially different, in the following sense: There exists a (homogeneous) linear mapping $\mathfrak{T}$ of $T(P)$ onto $T(Q)$ such that $\mathfrak{A} = \mathfrak{T}^{-1}\,\mathfrak{B}\mathfrak{T}$ ranges over all of $H(P)$ if $\mathfrak{B}$ ranges over all of $H(Q)$, and vice versa. (Thus the two groups are not only isomorphic, but are even *conjugates* of each other within the full linear group.) [*Hint*. Let V be a path from P to Q and let $\mathfrak{T}$ be the mapping of $T(P)$ onto $T(Q)$ induced by parallel displacement along V. Now, to each path W that begins and ends at Q, assign the path $V^{-1}\,WV$ that begins and ends at P.]

10.2 Geometric Interpretation of the Curvature Tensor

We shall consider special paths W_ε that are somewhat like parallelograms, with one vertex at a given point P. The coordinate system will be chosen in such a way that the coordinates of P are all zero and that $\Lambda^i_{jk}(0) = 0$; the latter is possible, by §9.4. For fixed, linearly independent η^i, ζ^i, we define the path W_ε as follows (see Fig. 30):

$$x^i(t) = \begin{cases} t\varepsilon\eta^i & \text{for} \quad 0 \leqq t \leqq 1, \\ \varepsilon\eta^i + (t-1)\,\varepsilon\,\zeta^i & \text{for} \quad 1 \leqq t \leqq 2, \\ \varepsilon\,(\eta^i + \zeta^i) - (t-2)\,\varepsilon\,\eta^i & \text{for} \quad 2 \leqq t \leqq 3, \\ \varepsilon\,\zeta^i\,(1 - (t-3)) & \text{for} \quad 3 \leqq t \leqq 4. \end{cases}$$

Now let ξ^i be a vector at P which is to undergo parallel displacement along the path $x^i(t)$, such that the vector assigned to it under that displacement at $x^i(t)$ is $\xi^i(t)$. The positive parameter ε will be allowed, subsequently, to tend to zero. By Taylor's theorem,

$$\xi^i(t+1) = \xi^i(t) + \frac{d\xi^i(t)}{dt} + \frac{1}{2}\frac{d^2\xi^i(t)}{dt^2} + \cdots .$$

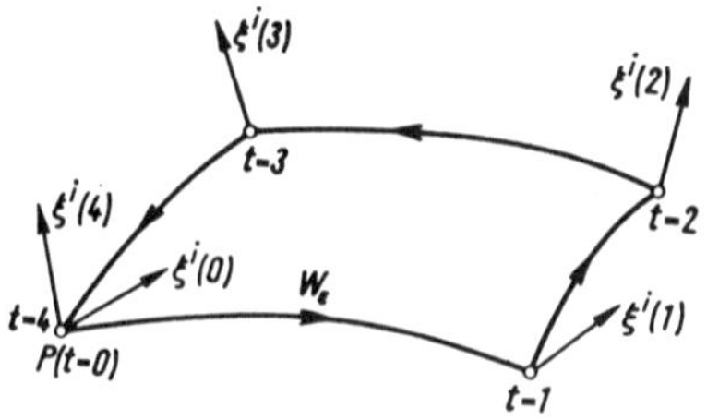

FIG. 30. Geometric interpretation of the curvature tensor.

Here, on account of parallelism, we have

$$\frac{d\xi^i(t)}{dt} = -\,\Lambda^i_{jk}\,\xi^k\,\dot{x}^j$$

$$\frac{d^2\xi^i}{dt^2} = -\,\Lambda^i_{jk,l}\,\xi^k\,\dot{x}^j\,\dot{x}^l + \Lambda^i_{jk}\,\Lambda^k_{rs}\,\xi^r\,\dot{x}^j\,\dot{x}^s,$$

where all quantities are to be evaluated at t. Let us now concentrate on quantities of order up to 2 in ε. We first use a Taylor expansion also for the components of the affine connection:

$$\Lambda^i_{jk}(t) = 0 + \Lambda^i_{jk,l}\,\Delta x^l + \cdots .$$

Towards computing the change in the vector ξ^i due to parallel displacement along the closed path W_ε, we note that

$$\xi^i(4) - \xi^i(0) = \xi^i(4) - \xi^i(3) + \xi^i(3) - \xi^i(2) + \xi^i(2) - \xi^i(1) + \xi^i(1) - \xi^i(0)$$

and

$$\frac{d\xi^i(0)}{dt} = 0$$

$$\frac{d\xi^i(1)}{dt} = -\varepsilon^2 \Lambda^i_{jk,l}(0)\, \xi^k \zeta^j \eta^l, \qquad \frac{d\xi^i(2)}{dt} = +\varepsilon^2 \Lambda^i_{jk,l}\, \xi^k \eta^j (\eta^l + \zeta^l)$$

$$\frac{d\xi^i(3)}{dt} = +\varepsilon^2 \Lambda^i_{jk,l}\, \xi^k \zeta^j \zeta^l$$

and

$$\frac{d^2\xi^i(0)}{dt^2} = -\varepsilon^2 \Lambda^i_{jk,l}\, \xi^k \eta^j \eta^l = \frac{d^2\xi^i(2)}{dt^2}$$

$$\frac{d^2\xi^i(1)}{dt^2} = -\varepsilon^2 \Lambda^i_{jk,l}\, \xi^k \zeta^j \zeta^l = \frac{d^2\xi^i(3)}{dt^2},$$

where, in all equations, terms in ε^3 (and higher powers) have been neglected and all functions are to be evaluated at 0.

We thus obtain

$$\frac{1}{\varepsilon^2}(\xi^i(4) - \xi^i(0)) = \Lambda^i_{jk,l}(\eta^j \zeta^l - \eta^l \zeta^j) + \varepsilon(\ldots) = (\Lambda^i_{jk,l} - \Lambda^i_{lk,j})\, \xi^k \eta^j \zeta^l + \varepsilon(\ldots).$$

Now, since $\Lambda^i_{jk}(0) = 0$ holds in the special coordinate system we are using, it follows that the difference of the derivatives of Λ^i_{jk} on the right-hand side is just the curvature tensor, whence

$$\frac{1}{\varepsilon^2}(\xi^i(4) - \xi^i(0)) = R^i{}_{kjl}\, \xi^k \eta^j \zeta^l + \varepsilon(\ldots).$$

Hence the mapping $\mathfrak{A}(W_\varepsilon)$ associated with our "parallelogram" path in accordance with §10.1 is given by

$$A^i_k = \delta^i_k + \varepsilon^2 R^i{}_{kjl} \eta^j \zeta^l + \varepsilon^3(\ldots).$$

This shows that the smaller ε is, the better does the term involving the curvature tensor approximate the difference $\xi^i(4) - \xi^i(0)$, and we can restate this in the following way: The curvature tensor is a measure of the change in a vector under parallel displacements around "infinitesimal" parallelograms. This interpretation, being invariant, is also independent of the special coordinate system we have used.

10.3 The (Locally) Affine Spaces as Spaces with Vanishing Curvature Tensor

We now turn to the question of determining the conditions under which the curvature tensor of a space with affine connection vanishes identically. In the theory of surfaces this happens, as we know, in the case of developable surfaces and in that case only, that is, in the case of surfaces whose intrinsic geometric properties are like those of the Euclidean plane. For dimensionality n, we know that the curvature tensor of n-dimensional Euclidean space vanishes identically; for a parallel vector field along an arbitrary curve, with t as the parameter, we have $d\xi^i/dt = 0$ (in Cartesian coordinates), so that $\Lambda^i_{jk} \equiv 0$ and, therefore, certainly $R^i{}_{hjk} = 0$. The following result is in the converse direction:

Theorem 10.3.1. *If $R^i{}_{hjk} = 0$, then, in some neighborhood U of the point P_0, it is possible to introduce a coordinate system $\bar{x}$ such that $\bar{\Lambda}^i_{jk} = 0$ in U, so that parallel vectors $\bar{\xi}^i(\bar{x})$, $\bar{\xi}^i(\bar{x}_1)$ at any two points P, P_1 of U always have the same components. Thus parallel displacement within U is independent of the path.*

Proof. We first prove the following: If $R^i{}_{hjk} = 0$, then there exists a neighborhood U_1 of P_0 such that, for every vector ξ^i_0 at P_0, there is a vector field $\xi^i(x)$, defined in U_1, consisting of parallel vectors (independent of the path) and satisfying $\xi^i(x_0) = \xi^i_0$. This assertion is equivalent to the statement that the system of differential equations

$$\frac{\partial \xi^i}{\partial x^k} = - \Lambda^i_{hk}\, \xi^h$$

has a solution satisfying $\xi^i(x_0) = \xi^i_0$, and this statement is seen to be true if we observe that the relevant integrability condition is precisely that $R^i{}_{hjk} = 0$ (see Appendix, II.2). This already establishes the independence of the path for parallel displacement in a neighborhood of P_0.

Among these parallel vector fields, we now select those n fields for which

$$\xi^i_{(k)}(x_0) = \delta^i_k \qquad (k = 1, \ldots, n). \tag{10.1}$$

On account of continuity,

$$\det(\xi^i_{(k)}) \neq 0 \tag{10.2}$$

holds throughout a neighborhood U_2 of P_0. Now the system

$$\frac{\partial x^i}{\partial \bar{x}^k} = \xi^i_{(k)}(x) \tag{10.3}$$

satisfies the integrability condition, on account of

$$\frac{\partial^2 x^i}{\partial \bar{x}^k\, \partial \bar{x}^j} = \frac{\partial \xi^i_{(k)}}{\partial x^l}\, \xi^l_{(j)} = - \Lambda^i_{hl}\, \xi^h_{(k)}\, \xi^l_{(j)}$$

and because the Λ^i_h are symmetric. Therefore there is a neighborhood U of P_0 in which solutions of (10.3) exist, and these solutions possess inverses $\bar{x}^k(x^i)$ because of (10.2). These inverse functions define an admissible coordinate transformation and, in the new coordinates, our n vector fields satisfy

$$\frac{\partial x^l}{\partial \bar{x}^i} \bar{\xi}^i_{(k)} = \xi^l_{(k)},$$

hence, by (10.3),

$$\xi^l_{(i)} \bar{\xi}^i_{(k)} = \xi^l_{(k)},$$

so that, by (10.2),

$$\bar{\xi}^i_{(k)} = \delta^i_k$$

in all of U. Since we are dealing with parallel vector fields, it follows that

$$0 = \bar{\xi}^i_{(k),j} + \bar{\Lambda}^i_{jr} \bar{\xi}^r_{(k)} = 0 + \bar{\Lambda}^i_{jr} \delta^r_k = \bar{\Lambda}^i_{jk}.$$

Hence in the new coordinate system, the components of the affine connection vanish in all of U, and the theorem is proved.

Note that, in general, the differentiability class of the new coordinate system is lower than that of the original coordinates. In contrast, the components of the affine connection are even analytic in the new coordinate system, since they vanish identically.

We see that locally, at least, the behavior of a space with affine connection whose curvature tensor vanishes is like that of affine space: In suitable coordinates, the corresponding components of parallel vectors are equal and parallel displacement is independent of the path. The restriction "locally" (or "in the small") cannot, in general, be dispensed with, as is shown by examples [such as the affine torus obtained from an affine (x, y)-plane by identifying points whose coordinates are congruent modulo 1].

Exercise 10.3.2. *Prove*: For $R^i{}_{hjk} = 0$ to hold in U, it is necessary and sufficient that $H(P_0)$ — for paths restricted to U — should consist of its identity element only.

10.4 The Auto-Parallels of a Connection

There is a close connection between geodesic parallelism and the geodesic curves on a surface: The tangent vectors of a geodesic curve are related by parallel displacement along the curve (cf. §4.7). We are accordingly led to the following generalization:

Definition. *If a curve, in a space with a linear connection, is such that all of its tangent vectors are related by parallel displacement along the curve, then the curve is called an auto-parallel curve (or, for short, an auto-parallel) of the connection.* (geodesic = auto-parallel)

It follows that an auto-parallel $x(t)$ satisfies the system of differential equations $\ddot{x}^i + \Lambda^i_{jk}\,\dot{x}^j\,\dot{x}^k = 0$. This shows that the auto-parallels of the linear connection with components Λ^i_{jk} are the same as those of the affine connection with components $\frac{1}{2}(\Lambda^i_{jk} + \Lambda^i_{kj})$. We may therefore confine our attention to affine connections and we shall thus again assume, in the next section and in the remainder of the present one, that $\Lambda^i_{jk} = \Lambda^i_{kj}$.

Concerning existence and uniqueness of auto-parallels, we have the following result:

Theorem 10.4.1. *Let P_0 be a point of a space with an affine connection. Then for every direction at P_0, there is an auto-parallel through P_0 having this direction at P_0 and this auto-parallel is uniquely determined throughout some neighborhood of P_0.*

Proof. First, the system of differential equations

$$\ddot{x}^i + \Lambda^i_{jk}\,\dot{x}^j\,\dot{x}^k = 0 \tag{10.4}$$

has, for t sufficiently small, exactly one solution $x^i(t)$ such that $x^i(t_0) = \underset{0}{x}^i$, $\dot{x}^i(t_0) = \underset{0}{\xi}^i$. It only remains to show that the initial conditions $x^i(t_1) = \underset{0}{x}^i$, $\dot{x}^i(t_1) = \lambda \cdot \underset{0}{\xi}^i$ lead to the same solution curve. Indeed,

$$x^i(t) = x^i(\lambda(t - t_1) + t_0)$$

is a solution that corresponds to these new initial conditions, and, by the uniqueness theorem for such systems of differential equations, it is the only one. Thus, the new solution curve differs from the previous one only in regard to a different choice of parameter and the theorem is proved.

Let us carry out an arbitrary parameter transformation $t = t(\tau)$ for an auto-parallel curve. Under it, the system (10.4) of differential equations is transformed into

$$\frac{d^2x^i}{d\tau^2} + \Lambda^i_{jk}\frac{dx^j}{d\tau}\frac{dx^k}{d\tau} = -\frac{dx^i}{d\tau}\frac{d^2\tau}{dt^2}\bigg/\left(\frac{d\tau}{dt}\right)^2. \tag{10.5}$$

If a parameter τ is such that the right-hand side of (10.5) vanishes — so that the transformed differential equations are again of the form (10.4) —

then τ is called a *natural parameter* of the auto-parallel. The admissibility of a parameter transformation is subject to the same definition as in the elementary theory of curves (cf. I, §1), so that $dx^i/d\tau \neq 0$ holds; hence (10.5) shows that natural parameters are characterized by $d^2\tau/dt^2 = 0$, which proves the following result:

Theorem 10.4.2. *If t is a natural parameter on an auto-parallel, then all other natural parameters for the curve are given by*

$$\tau = c_0 t + c_1, \qquad c_i = \text{const.}, \qquad c_0 \neq 0.$$

Thus, for geodesic curves on a surface, for instance, all natural parametrizations are obtained from the arc length parameter s by linear parameter transformations $c_0 s + c_1$.

We have also the following result:

Theorem 10.4.3. *Any solution curve of*

$$\frac{d^2x^i}{d\tau^2} + \Lambda^i_{jk}\frac{dx^j}{d\tau}\frac{dx^k}{d\tau} = -\mu(\tau)\frac{dx^i}{d\tau}$$

is an auto-parallel of the affine connection whose components are the Λ^i_{jk}.

For, considering (10.5), we merely have to introduce by means of

$$\frac{d^2\tau}{dt^2} = \mu(\tau)\left(\frac{d\tau}{dt}\right)^2$$

a natural parameter t.

10.5 Projective Equivalence of Affine Connections

The following question may now be raised: Assuming that the net of all the auto-parallel curves of an affine connection is assigned, is the affine connection itself uniquely determined thereby? The answer is in the negative; in fact, Weyl (1921) proved the following theorem:

Theorem 10.5.1. *Two affine connections with respective components $\tilde{\Lambda}^i_{jk}$, Λ^i_{jk} have the same auto-parallels if and only if there exists a covariant vector φ_j such that*

$$A^i_{jk} \underset{\text{def}}{=} \tilde{\Lambda}^i_{jk} - \Lambda^i_{jk} = \delta^i_j \varphi_k + \delta^i_k \varphi_j.$$

Proof. The condition is sufficient, for it implies that

$$\tilde{\Lambda}^i_{jk}\dot{x}^j\dot{x}^k = \Lambda^i_{jk}\dot{x}^j\dot{x}^k + 2\dot{x}^i\varphi_k\dot{x}^k,$$

so that we can use $\mu = 2\varphi_k \dot{x}^k$ to introduce a new parameter σ such that, as Theorem 10.4.3 shows, not only $\ddot{x}^i + \bar{\Lambda}^i_{jk} \dot{x}^j \dot{x}^k = 0$ holds, but also

$$\frac{d^2x^i}{d\sigma^2} + \Lambda^i_{jk} \frac{dx^j}{d\sigma} \frac{dx^k}{d\sigma} = 0.$$

To prove the necessity of the condition, let $x^i(t)$ be an auto-parallel for both of the given connections. Then we have, say,

$$\ddot{x}^i + \bar{\Lambda}^i_{jk} \dot{x}^j \dot{x}^k = 0$$

and

$$\ddot{x}^i + \Lambda^i_{jk} \dot{x}^j \dot{x}^k = -\lambda \dot{x}^i,$$

whence, by subtraction,

$$A^i_{jk} \dot{x}^j \dot{x}^k = \lambda \dot{x}^i,$$

and therefore

$$(\dot{x}^r A^i_{jk} - \dot{x}^i A^r_{jk}) \dot{x}^j \dot{x}^k = 0.$$

Since this must be valid for all directions, the cubic form on the left vanishes identically and, therefore, all of its coefficients must be zero. Hence,

$$0 = \delta^r_s A^i_{jk} - \delta^i_s A^r_{jk} + \delta^r_j A^i_{ks} - \delta^i_j A^r_{ks} + \delta^r_k A^i_{sj} - \delta^i_k A^r_{sj}.$$

Contraction over r and s yields

$$0 = (n+1) A^i_{jk} - \delta^i_j A^r_{kr} - \delta^i_k A^r_{jr}.$$

If we set $\varphi_k = [1/(n+1)]A^r_{kr}$, we see that the proof will be complete if we can show that φ_k is a covariant vector. But this follows from the fact that A^i_{jk} is a tensor. [Form the difference of the transformation formulas, (9.4).]

Two affine connections that have the same auto-parallels are called *projectively equivalent.* This terminology is motivated by the fact that a projective mapping of ordinary space is a mapping that takes straight lines into straight lines, and that the role of straight lines is played by the auto-parallels in the present context.

10.6 General Remarks Concerning Terminology

In differential geometry, one must distinguish, for geometric properties, between validity in the whole space, validity in a sufficiently small neighborhood of every point, and validity in the tangent space only. The necessity of such distinctions is illustrated by the attribute "affine" whose use implies that a concept of parallelism of vectors is in force. There are spaces — namely, the affine spaces proper — in which parallelism is defined for all vectors throughout the space, that is, "globally" (or "in the large"). There are other spaces in which one can speak of parallelism of vectors only in sufficiently small neighborhoods of a point, thus only "locally"; these spaces are characterized by the vanishing of the curvature tensor. Finally, for every point of a differentiable manifold, the tangent space at the point is a vector space and, therefore, an affine space and we can therefore say that every differentiable manifold is "pointwise" affine. Furthermore, an intermediate position between the last two possibilities is occupied by the spaces with affine connection, where parallelism is defined in terms of a differential relation. In more old-fashioned terminology, this used to be expressed by saying that parallelism is defined for vectors at infinitesimally neighboring points (x) and $(x + dx)$; thus the spaces with affine connection are, so to speak, "infinitesimally" affine.

For differentiable manifolds, we shall therefore have to differentiate four types of properties. A property is called

global, if it is applicable to the whole manifold;

local, if every point has a neighborhood in which the property holds;

infinitesimal, if it is defined in terms of differential equations;

pointwise, if it applies to individual points and their tangent spaces.

The dimensionality of a space is, accordingly, a pointwise property. Local compactness is a local property: For every point, there exists a neighborhood such that addition of all of its points of accumulation makes it compact (in the sense of every infinite sequence possessing a convergent subsequence). An important global property defined for spaces with a linear connection is the property of *completeness*: Such a space is called complete if every auto-parallel can be extended indefinitely in both of its directions. (Put more precisely: If t is a natural parameter on an auto-parallel piece of curve, then the space contains an auto-parallel defined over all of $-\infty < t < +\infty$ and containing the given piece of curve.) The whole plane and the sphere are examples of complete surfaces. The open northern hemisphere is also a manifold, to be sure, but it is not complete.[1] Another global property defined for manifolds is compactness.

Theorems that allow, say, for local or global conclusions to be drawn from assumptions of the infinitesimal type, are always of particular interest.

[1] Equivalent conditions for completeness have been investigated especially by H. Hopf and W. Rinow: *Comment. Math. Helv.* **3**, 209–224 (1931).

We had an example of such a theorem in the result that the locally affine character of a space is a consequence of the vanishing of the curvature tensor.

§ 11. Foundations of Riemannian Geometry

11.1 The Riemannian Metric

We wish to introduce in a differentiable manifold the possibility of measuring the length of a curve, and to this end we assume that there is defined on the manifold a doubly covariant, symmetric tensor $g_{ik}(x)$. In terms of it, we define the length of a piece of curve $x^i(t)$, $t_0 \leqq t \leqq t_1$, to be

$$s(t_1) - s(t_0) = \int_{t_0}^{t_1} \sqrt{g_{ik}\, \dot{x}^i\, \dot{x}^k}\, dt. \tag{11.1}$$

A metric of this sort is called *Riemannian* and a space with such a metric is called a *Riemannian space* V_n. Examples of two-dimensional Riemannian spaces V_2 are furnished by the surfaces in Euclidean space. Indeed, Riemann (1854) developed his geometries with the Gaussian theory of surfaces in mind. He aimed at developing the intrinsic geometry of surfaces independently of any imbedding, in such a way that the entities of intrinsic geometry should be founded on the metric alone; in particular, this would apply also to the Gaussian curvature, which the *theorema egregium* shows to belong to the intrinsic geometry of the surface. In this program, Riemann also allowed for a generalization as regards dimensionality (dimension n instead of just 2), and this has given rise to a plethora of applications outside of surface theory also.

We want the radicand in (11.1) always to be positive and to have any finite piece of curve to be of positive length. To this end, we shall presuppose that $g_{ik}\, \xi^i\, \xi^k > 0$ for $\xi^i \neq 0$, an assumption that reduces to a valid fact in the case of the theory of surfaces. This implies, in particular, that $g = \det(g_{ik}) > 0$. [Many of the results that we shall prove in this section remain valid also in the more general case of indefinite g_{ik}, with $\det(g_{ik}) \neq 0$, which is important for the theory of relativity. We shall, however, confine ourselves to the definite case only, which is of greater interest from the geometric point of view.] We shall also have to impose differentiability conditions. In general, we shall find it sufficient to require that $g_{ik}(x)$ be of class C^3 and that the manifold itself be of class C^4.

Since g_{ik} is a tensor, $g_{ik}\dot{x}^i\dot{x}^k$ is an invariant and the arc length (11.1) is therefore independent of the coordinate system. It is also independent of the particular parametrization of the curve, as the reader can easily verify for himself. The tensor g_{ik} is called the (*fundamental*) *metric tensor* or the *fundamental covariant tensor* of the Riemannian space.

It may appear somewhat unsatisfying to introduce the Riemannian metric merely as a formal generalization of the metric on a surface. The following approach might be more satisfactory from a geometric standpoint. Assume that a measurement of lengths is defined in the space having the following properties: There is assigned to every smooth curve C a positive number $L(C)$ as its length, in such a way that, if C consists of the pieces C_1 and C_2, then $L(C) = L(C_1) + L(C_2)$. Now, there are certainly non-Riemannian metrics with these properties, for example,

$$L(C) = \int_{t_0}^{t_1} (|\dot{x}^1| + |\dot{x}^2|)\, dt$$

or

$$L(C) = \int_{t_0}^{t_1} \sqrt[4]{(\dot{x}^1)^4 + (\dot{x}^2)^4}\, dt.$$

We can raise the question whether, among these more general metrics, the Riemannian metrics are characterized by any particular geometric properties. This question will be answered later, in §15 below, where we shall deal with the more general type of metric and with the special place that the Riemannian metrics occupy. We shall first turn our attention to the geometry of the Riemannian spaces themselves.

We define the length of a contravariant vector ξ^i at the point (x) to be the number $\sqrt{g_{ik}(x)\xi^i\xi^k}$ and we define the angle φ between two vectors ξ^i and η^i at (x) by means of

$$\cos\varphi = \frac{g_{ik}\,\xi^i\eta^k}{\sqrt{g_{ik}\,\xi^i\xi^k}\sqrt{g_{ik}\,\eta^i\eta^k}}\,.$$

The angle between two curves that intersect at a point is defined as being the angle between their tangent vectors at the point.

Exercise 11.1.1. *Surfaces of Dimension p in Euclidean n-space as Riemannian Spaces of Dimension p.* Let such a surface be given by $\mathfrak{x} = \mathfrak{x}(u^1, \ldots, u^p) = \mathfrak{x}(u^i)$. A curve $\mathfrak{x}(t) = \mathfrak{x}(u^i(t))$ on the surface has arc length

$$s(t) - s(t_0) = \int_{t_0}^{t} \sqrt{\dot{\mathfrak{x}}^2}\, dt = \int_{t_0}^{t} \sqrt{g_{ik}\, \dot{u}^i\, \dot{u}^k}\, dt,$$

where

$$g_{ik} = \mathfrak{x}_i\, \mathfrak{x}_k, \qquad \mathfrak{x}_i \underset{\text{def}}{=} \frac{\partial \mathfrak{x}}{\partial u^i}.$$

Exercise 11.1.2. *Euclidean Space in Arbitrary Curvilinear Coordinates.* If we pass from rectangular Cartesian coordinates x^i of a Euclidean space to arbitrary curvilinear coordinates by means of a differentiable coordinate transformation $x^i = x^i(\bar{x}^k)$, the fundamental metric tensor becomes

$$\bar{g}_{ik}(\bar{x}) = \delta_{rs} x^r_{,i}\, x^s_{,k}.$$

The associated quadratic form is positive-definite, provided that the Jacobian (functional determinant) of the transformation is not zero.

Supplementary Remarks Concerning Imbedding Possibilities. Every p-dimensional surface in Euclidean n-space carries an induced Riemannian metric. One may ask if, conversely, every Riemannian metric can be realized in terms of such a surface metric, i.e., if every Riemannian space of dimension p can be imbedded isometrically in a Euclidean n-space. This is a problem of analysis; the question is whether or not the system $\mathfrak{x}_i\, \mathfrak{x}_k = g_{ik}$ of partial differential equations is solvable. We have to confine ourselves here to references to a few results.

(a) *Local imbedding,* referring to the question as to whether sufficiently small portions of a Riemannian space can be imbedded: An old conjecture of Schläfli (1871) was proved by M. Janet (1926) and É. Cartan (1927). It says that every p-dimensional Riemannian space (of sufficiently high class C^m, say $m \geqq 4$) can be imbedded isometrically in Euclidean n-space if $n \geqq \binom{p+1}{2}$. In particular, every two-dimensional Riemannian space can be realized locally as a surface in $n = 3$.

(b) *Global Imbedding,* referring to the question whether an entire Riemannian manifold can be imbedded. Here, the situation is governed by less favorable results, and some of its aspects still await clarification. At any rate, imbeddings of sufficiently high differentiability class always exist for $n \geqq 2p$; on the other hand, not every two-dimensional Riemannian space can be globally imbedded, in an isometric way, into three-dimensional Euclidean space if sufficient differentiability conditions are imposed. (According to D. Hilbert — see his "Grundlagen der Geometrie," 8th ed., Appendix V. Stuttgart, 1956 — there does not exist any class C^4-imbedding of the hyperbolic plane in three-dimensional Euclidean space; there do exist class C^1-imbeddings, but these are not satisfactory from the differential-geometric point of view, since even the calculation of the second fundamental form requires differentiability of the second order.) Recent results and bibliographical references concerning questions of imbedding can be found, e.g., in N. H. Kuiper, *Indagationes Math.* **17**, 545 and 683 (1955); K. Leichtweiss, *Math. Ann.* **130**, 442 (1956); J. Nash, *Ann. of Math.* **63**, 20 (1956).

11.2 The Affine Connection Associated with a Riemannian Metric

For the further development of Riemannian geometry, the introduction of a certain covariant differentiation process turns out to be indispensable. Since a parallel displacement is associated with this, we may reasonably hope for a choice of differentiation process such that it will lead to a generalization of the geodesic parallelism in the theory of surfaces.

We shall call an affine connection *compatible* with a Riemannian metric if the lengths of parallel vectors are the same under the associated infinitesimal parallel displacement. Thus, for instance, the geodesic parallelism in the theory of surfaces is compatible with the metric of the surface, as is shown by Theorem 4.7.3. Now we have the following result:

Theorem 11.2.1. *In a Riemannian space with metric tensor* g_{ik}, *there exists one and only one affine connection that is compatible with the metric. The components of this connection are*

$$\Lambda^i_{jk} = \Gamma^i_{jk} \underset{\text{def}}{=} \tfrac{1}{2} g^{ir}(g_{jr,k} - g_{jk,r} + g_{rk,j}). \tag{11.2}$$

Here too, we shall call the quantities Γ^i_{jk} the *Christoffel symbols*. Theorem 11.2.1 gives not only a geometrically sensible definition of a covariant differentiation but, at the same time, an intrinsic geometric characterization of the geodesic parallelism of the theory of surfaces.

Proof. The proof that (11.2) is sufficient for the preservation of lengths (and also of angles), under the associated parallel displacement, is exactly like the proof of sufficiency for Theorem 4.7.3. It only remains to prove the necessity of the condition. Assume, then, that $x^i(t)$ is any curve and that, for a parallel vector field $\xi^i(t)$, we have

$$0 = \frac{d}{dt}(g_{ik}\,\xi^i\,\xi^k) = g_{ik,l}\,\dot{x}^l\,\xi^i\,\xi^k + g_{ik}\frac{d\xi^i}{dt}\,\xi^k + g_{ik}\frac{d\xi^k}{dt}\,\xi^i$$

$$= (g_{ik,l} - g_{rk}\Lambda^r_{li} - g_{ir}\Lambda^r_{lk})\,\dot{x}^l\,\xi^i\,\xi^k.$$

Since this holds at every point for all x and ξ and since the expression is symmetric in i and k, it follows that

$$g_{ik,l} = g_{rk}\Lambda^r_{li} + g_{ir}\Lambda^r_{lk} \tag{11.3}$$

and from this, after use of the symmetry of the g_{ik} and Λ^i_{jk}, that

$$g_{il,k} - g_{ik,l} + g_{lk,i} = 2\,g_{rl}\Lambda^r_{ik},$$

and now we can obtain the asserted condition by inner multiplication with $\frac{1}{2}g^{lj}$.

As by-products of the proof, we have also obtained

Corollary 1. *Under the infinitesimal parallel displacement associated with Γ^i_{jk}, angles are preserved.*

Corollary 2. (the so-called lemma of Ricci). *The covariant differentiation given by Γ^i_{jk} satisfies $g_{ik;l} = 0$.*

[The proof follows from (11.3) and from formula (E) of §9.5 for the covariant differentiation of covariant tensors of order 2.]

From now on, and without mentioning it each time explicitly, we shall always use in Riemannian spaces the special covariant differentiation introduced above. The associated parallel displacement will be referred to as geodesic parallelism, just as in the theory of surfaces.

What statements can be made concerning the associated holonomy group? Let us first carry out a linear transformation of the coordinates which is such that at the point P_0 in question, $g_{ik} = \delta_{ik}$ holds in the new coordinate system. This is always possible, because the quadratic form $g_{ik}\,\xi^i\,\xi^k$ is definite. If we now carry out a geodesic parallel displacement along any closed path that begins and ends at P_0, we obtain a length-preserving linear mapping of the tangent space at P_0 onto itself. But a linear transformation that leaves the value of $\delta_{ik}\,\xi^i\,\xi^k$ unchanged is an orthogonal transformation, hence a rotation or a reflection or a combination of such. This proves

Theorem 11.2.2. *With a suitable choice of coordinates, the holonomy group at any point of a Riemannian space is an orthogonal group.*

11.3 Geodesic Curves

The auto-parallels of geodesic parallelism are the solution curves of the differential equation

$$\ddot{x}^i + \Gamma^i_{jk}\,\dot{x}^j\,\dot{x}^k = 0. \tag{11.4}$$

On the other hand, we can prove in literally the same way as for the corresponding theorem (Theorem 4.4.1) of the theory of surfaces that the curve of minimum length between two assigned points must satisfy the same system (11.4) of differential equations. Calling such extremal curves *geodesic curves*, we therefore have

Theorem 11.3.1. *The auto-parallels of geodesic parallelism are, at the same time, the geodesic curves of the Riemannian metric.*

With the help of this theorem, we can establish the following result: *If t is a natural parameter on a geodesic curve and if s is arc length, then* $t = c_0 s + c_1$.

For, the proof of Theorem 11.3.1 shows that s is a natural parameter and, by Theorem 10.4.2, every other natural parameter is obtained from one of them by a linear parameter transformation.

In what precedes we have not proved the fact that for sufficiently small portions of the space (that is, locally), the geodesic curves actually do furnish the shortest connections between any two of their points. The proof requires the introduction of special coordinates, just as in the theory of surfaces (see Theorem 4.6.1).

11.4 Theory of Curves

Let $x^i(s)$ be any curve that is sufficiently differentiable in a Riemannian space. Without loss of generality, we assume the curve to be parametrized in terms of its arc length, so that

$$\tilde{g}_{ik}(x(s))\, x^{i'}\, x^{k'} = 1. \tag{11.5}$$

We now take the theory of curves in Euclidean 3-space, developed in Chapter I, as a model in accordance with which we want to determine, for our curve in the Riemannian space, its curvature as well as a moving frame of vectors that will play the role of the moving trihedron of Frenet. In Euclidean space, we obtained the principal normal vector by differentiating the tangent vector $x^{i'}(s)$, and the length of the derived vector was the curvature $\kappa(s)$ of the curve. For present purposes, we replace ordinary by covariant differentiation; from the tangent vector

$$\xi^i_{(1)} \underset{\text{def}}{=} \frac{dx^i}{ds} \tag{11.6}$$

we obtain in this way

$$\xi^i_{(2)} \underset{\text{def}}{=} \frac{D\xi^i_{(1)}}{ds} = \frac{d\xi^i_{(1)}}{ds} + \Gamma^i_{jk}\, \xi^k_{(1)}\, \frac{dx^j}{ds} = x^{i''} + \Gamma^i_{jk}\, x^{j'}\, x^{k'}. \tag{11.7}$$

Using Ricci's lemma of §11.2 (see Corollary 2), we obtain from (11.5) by covariant differentiation that

$$g_{ik}\, \xi^i_{(1)}\, \xi^k_{(2)} = 0. \tag{11.8}$$

Hence the vector (11.7) is orthogonal to the tangent of the curve; we call its direction the *principal normal direction*. If $\eta^i_{(2)}$ is the corresponding unit vector, then

$$\xi^i_{(2)} = \kappa_{(1)} \eta^i_{(2)}, \qquad \kappa_{(1)} \geqq 0, \qquad g_{ik} \eta^i_{(2)} \eta^k_{(2)} = 1, \tag{11.9}$$

and — following again the model of the theory of curves in Euclidean 3-space — we call $\kappa_{(1)}$ the *curvature* and $\eta^i_{(2)}$ the *principal normal vector*.

(We note that the principal normal direction is determined only if $\xi^i_{(2)} \neq 0$.) The above formulas show that for the case of surfaces in 3-space, $\kappa_{(1)}$ coincides with the geodesic curvature κ_g, since, according to Eq. (4.1), the latter is precisely the length of the vector whose components are given by (11.7). But this means that we now possess a definition of geodesic curvature that is independent of imbedding in space, namely, as the length of the covariant derivative of the unit tangent vector. More generally, (11.7) shows that in arbitrary Riemannian spaces, the geodesic curves are characterized by $\kappa_{(1)} = 0$.

Let us now generalize the Frenet formulas to the case of Riemannian spaces. To this end, we assume that $x^i(s)$ is of class C^{n+1}, and if the n vectors $\xi^i_{(k)}$ are defined by

$$\xi^i_{(k)} = \frac{D}{ds} \xi^i_{(k-1)} \qquad \text{for} \qquad 2 \leqq k \leqq n$$

$$\xi^i_{(1)} = \frac{dx^i}{ds},$$

we further assume that these vectors are linearly independent. We can then construct from them the n vectors $\eta^i_{(k)}$ that result from applying E. Schmidt's orthogonalization process (cf. §8.4). This implies that $\eta^i_{(k)}$ is a linear combination of $\xi^i_{(1)}, \ldots, \xi^i_{(k)}$, so that the covariant derivative of $\eta^i_{(k)}$ is a linear combination of $\xi^i_{(1)}, \ldots, \xi^i_{(k+1)}$. Since each $\xi^i_{(j)}$ can be expressed in terms of the $\eta^i_{(r)}$ with $r \leqq j$, we can set

$$\frac{D\eta^i_{(k)}}{ds} = \sum_{j=1}^{k+1} a_{kj} \eta^i_{(j)}.$$

Since the $\eta^i_{(k)}$ are an orthonormal set, we have $g_{ik} \eta^i_{(j)} \eta^k_{(l)} = \delta_{jl}$, whence it follows by covariant differentiation, again using Ricci's lemma, that

$$0 = g_{ik} \frac{D\eta^i_{(j)}}{ds} \eta^k_{(l)} + g_{ik} \eta^i_{(j)} \frac{D\eta^k_{(l)}}{ds} = a_{jl} + a_{lj}.$$

Hence the matrix of the a_{lj} is skew-symmetric and, since $j \geqq l + 2$ implies that $a_{lj} = 0$, it follows that the only places where the matrix can have nonzero entries are in the two diagonals next to the main diagonal. We write $\kappa_{(j)}$ for $a_{jl} = - a_{lj}$ if $l = j + 1$ and we then have the following Frenet formulas:

$$\frac{D\overset{i}{\eta}_{(1)}}{ds} = + \kappa_{(1)} \overset{i}{\eta}_{(2)}$$

$$\frac{D\overset{i}{\eta}_{(2)}}{ds} = - \kappa_{(1)} \overset{i}{\eta}_{(1)} + \kappa_{(2)} \overset{i}{\eta}_{(3)}$$

$$\cdots$$

$$\frac{D\overset{i}{\eta}_{(j)}}{ds} = - \kappa_{(j-1)} \overset{i}{\eta}_{(j-1)} + \kappa_{(j)} \overset{i}{\eta}_{(j+1)}$$

$$\cdots$$

$$\frac{D\overset{i}{\eta}_{(n)}}{ds} = - \kappa_{(n-1)} \overset{i}{\eta}_{(n-1)},$$

where the curvatures $\kappa_{(j)} = g_{ik} \dfrac{D\overset{i}{\eta}_{(j)}}{ds} \overset{k}{\eta}_{(j+1)}$ are invariants of the curve, since all other quantities occurring in this formula are tensors or vectors.

For $\Gamma^i_{jk} = 0$, that is, for the case of Euclidean space, and specializing furthermore to $n = 3$ or $n = 2$, we obtain precisely the Frenet formulas of Chapter I. For the case of surfaces ($n = 2$), the equations reduce to the first and last one and there is only one curvature, namely, the geodesic curvature $\kappa_g = \kappa_{(1)}$.

Exercise 11.4.1. *Prove*: Using the Gramian determinants $G_{(j)}$ of the vectors $\overset{i}{\xi}_{(j)}$, we can express the curvatures explicitly in terms of $x^i(s)$ and its derivatives, as follows:

$$\kappa_{(j)} = \frac{\sqrt{G_{(j-1)}\, G_{(j+1)}}}{G_{(j)}}$$

[*Hint.* Differentiate the defining equation for $\overset{i}{\eta}_{(j)}$ covariantly with respect to s, and use the determinant expressions given at the end of §8.4. The calculations are carried through in detail in W. Blaschke, *Math. Z.* **6**, 98 (1921).

Exercise 11.4.2. *The curvatures $\kappa_{(j)}$ constitute a complete system of invariants of a curve in Riemannian space.* Following the model of the proof of our earlier theorem for curves in Euclidean space (§2.3), using systematically covariant instead of ordinary differentiation, *prove* that for an assigned set of functions $\kappa_{(j)}(s)$ and an assigned initial frame of vectors, there exists one and only one curve with these assigned curvatures and initial vectors. (See Blaschke, *loc. cit.*, in Exercise 11.4.1.) In particular, therefore, the geodesic curvature is the only essential invariant of a curve on a surface, arc length being the curve parameter.

11.5 Riemannian Normal Coordinates

In Euclidean space, Cartesian coordinates play a distinguished role among all coordinate systems. Following Riemann, one can introduce similarly convenient coordinates in the more general case also. These so-called *normal coordinates* with center at a point P_0 have important properties in common with Cartesian coordinates.

Let P_0 be any point and let the coordinates x^i be geodesic at P_0, so that $\Gamma^i_{jk}(\underset{0}{x}) = 0$. In every direction ξ^i at P_0 (ξ^i: a unit vector), there is a geodesic curve, say $x^i(s)$, with $x^i(0) = \underset{0}{x}^i$, $x^{i\prime}(0) = \xi^i$. Using a Taylor expansion, we have

$$x^i(s) = \underset{0}{x}^i + s\, x^{i\prime}(0) + \frac{s^2}{2}\, x^{i\prime\prime}(0) + \frac{s^3}{3!}\, x^{i\prime\prime\prime}(0) + \ldots \tag{11.10}$$

or, using the differential equations for geodesic curves,

$$x^i(s) = \underset{0}{x}^i + s\, \xi^i - \frac{1}{3!}\, s^3\, \Gamma^i_{jk,l}(0)\, \xi^j\, \xi^k\, \xi^l + s^4(\ldots) \tag{11.11}$$

for all sufficiently small s. If we set $\bar{x}^i = s\xi^i$, then we have a coordinate transformation[1]

$$x^i = \underset{0}{x}^i + \bar{x}^i - \frac{1}{3!}\, \Gamma^i_{jk,l}(0)\, \bar{x}^j\, \bar{x}^k\, \bar{x}^l + \ldots,$$

defined in a neighborhood of P_0, since the Jacobian equals 1 at $x = \underset{0}{x}$ and is therefore not zero in some neighborhood of P_0. The differential equations for the geodesics in the new coordinates, $\bar{x}^{i\prime\prime} + \bar{\Gamma}^i_{jk}(\bar{x})\bar{x}^{j\prime}\bar{x}^{k\prime} = 0$, become, in view of $\bar{x}^i = \xi^i s$,

$$\bar{\Gamma}^i_{jk}\, \xi^j\, \xi^k = 0$$

or, upon multiplication by s^2,

$$\bar{\Gamma}^i_{jk}(\bar{x})\, \bar{x}^j\, \bar{x}^k = 0. \tag{11.12}$$

---

[1] The solutions of the differential equations for the geodesics are differentiable functions of the initial conditions ξ^i, which guarantees the required differentiability properties of the coordinate transformation. However, if the original coordinates were of class C^k, the normal coordinates will in general be of class C^{k-2} only.

Furthermore, the geodesics through P_0 are represented, in the new normal coordinates, by linear equations, $\bar{x}^i = \xi^i s$, just like the straight lines of Euclidean space in Cartesian coordinates. For other geodesics, not through P_0, nothing similar can be expected.

Exercise 11.5.1. *Prove*: If a coordinate system is a normal coordinate system simultaneously at all points of the space, then the space is locally Euclidean. [*Hint.* By (11.12), we have

$$\bar{\Gamma}^i_{jk}(\bar{x})(\bar{x}^j - \underset{0}{\bar{x}^j})(\bar{x}^k - \underset{0}{\bar{x}^k}) = 0$$

for all $\underset{0}{\bar{x}^j}$, from which it follows, on account of the symmetry of the Γ, that $\bar{\Gamma}^i_{jk} \equiv 0$, whence by (11.3), $\bar{g}_{ik} = \text{const.}$]

Exercise 11.5.2. *Geodesic Polar Coordinates for* $n = 2$. Given normal coordinates $\bar{x}$, one can pass from them by means of

$$\bar{x}^1 = u^1 \cos u^2, \qquad \bar{x}^2 = u^1 \sin u^2$$

to geodesic polar coordinates, where u^1 is the geodesic distance from P (the pole) and u^2 is the angle of the geodesic with a fixed direction. *Calculate* $d\bar{x}^i$ and *substitute* into the element of arc length $ds^2 = \bar{g}_{ik}\, d\bar{x}^i\, d\bar{x}^k$ in the normal coordinates; you will obtain, in the polar coordinates, $g_{11} = 1$, $g_{12} = 0$, and $g_{22} = (u^1)^2(\bar{g}_{11} \sin^2 u^2 - 2\bar{g}_{12} \sin u^2 \cos u^2 + \bar{g}_{22} \cos^2 u^2)$. Since, in the normal coordinates, the first derivatives of the $\bar{g}_{ik}$ vanish at P, it follows for the polar coordinates that

$$\lim_{u^1 \to 0} g_{22} = 0, \qquad \lim_{u^1 \to 0} (\sqrt{g_{22}})_{,1} = 1.$$

Since we know from Exercise 5.5.5 that in geodesic coordinates

$$\frac{\partial^2 f}{(\partial u^1)^2} = -Kf \qquad (f = \sqrt{g_{22}}),$$

it also follows that, for $u^1 \to 0$, we have $\partial^2 f/(\partial u^1)^2 \to 0$ and $\partial^3 f/(\partial u^1)^3 \to -K_0$, where K_0 is the Gaussian curvature at P. Thus we have the following Taylor expansions at P:

$$f = u^1 - \tfrac{1}{6} K_0 (u^1)^3 + (u^1)^4(\ldots) \qquad \text{and} \qquad g_{22} = (u^1)^2 - \tfrac{1}{3} K_0 (u^1)^4 + (u^1)^5(\ldots).$$

Exercise 11.5.3. From Exercise 11.5.2, it follows that the circumference U of a "geodesic circle" of radius $u^1 = r$ satisfies

$$U = \int_0^{2\pi} \sqrt{g_{22}}\, du^2 = 2\pi r - \tfrac{1}{3} K_0 \pi r^3 + r^4(\ldots)$$

and from this we obtain the celebrated formula of Bertrand and Puiseux (1848),

$$K_0 = \lim_{r \to 0} 3(2\pi r - U)/\pi r^3.$$

This means that the Gaussian curvature is obtainable in a very simple way from measurements of lengths within the surface, yielding, at the same time, a very "concrete" proof of the *theorema egregium*. (If $K_0 > 0$, then the circumference U is smaller than the Euclidean circumference $2\pi r$. This interpretation of the curvature as a kind of "shrinkage" is especially recommended to the attention of those philosophers who, time and again, take off from the term "curvature" and conclude that a "curved" space is unthinkable, unless imbedded in a Euclidean space!)

Exercise 11.5.4. *Prove* that, using the notation of Exercise 11.5.3 and, with F denoting the area of the circle in 11.5.3, we have

$$F = \int_0^{2\pi}\int_0^{r} f\, du^1\, du^2$$

$$K_0 = \lim_{r \to 0} \frac{12}{\pi} \frac{\pi r^2 - F}{r^4}.$$

11.6 The Riemannian Curvature of Space

Following Riemann, we can use the curvature tensor to introduce invariants that are much closer to the geometric significance of the curvature of a surface than is the curvature tensor itself. The latter, after all, was introduced in quite a formal manner in connection with the fact that second covariant derivatives are not independent of the order of the differentiations; and its only geometric meaning, so far, had to do with the change in direction of a vector under parallel displacement around infinitesimal parallelograms.

Let ξ^i, η^i be two vectors of the tangent space at P_0. Let us introduce normal coordinates with center at P_0 and in such a way that ξ^i and η^i are in the direction of the x^1-curves and of the x^2-curves at P_0, respectively. Consider the geodesic curves through P_0 whose tangent vectors at P_0 are of the form $\alpha\xi^i + \beta\eta^i$. In a neighborhood of P_0, these geodesics constitute a surface that, in our normal coordinates at P_0, is described by $\bar{x}^3 = \ldots = \bar{x}^n = 0$. Thus, the surface depends only on $\bar{x}^1$ and $\bar{x}^2$, and, by §5.5, its Gaussian curvature is given by

$$K = \frac{g_{2j}\, R^j{}_{112}}{g}$$

or, introducing the (purely covariant) curvature tensor

$$R_{ijkl} = g_{ir}\, R^r{}_{jkl} \tag{11.13}$$

and multiplying numerator and denominator by the only nonzero components of ξ and η,

$$K = -\frac{R_{2121}\,\xi^1\eta^2\xi^1\eta^2}{(g_{11}\,g_{22} - g_{12}\,g_{21})\,\xi^1\eta^2\xi^1\eta^2}. \tag{11.14}$$

We therefore have, in arbitrary coordinates,

$$K = +\frac{R_{ijkl}\,\eta^i\eta^k\xi^j\xi^l}{(g_{il}\,g_{jk} - g_{ik}\,g_{jl})\,\eta^i\eta^k\xi^j\xi^l}. \tag{11.15}$$

For, this expression is invariant and it reduces to (11.14) if we use the special (normal) coordinates. Thus the expression (11.15) is simply the Gaussian curvature of the two-dimensional surface spanned by the selected family of geodesics; it is called the (Riemannian) curvature of the space at the point (x) in the two-dimensional direction spanned by ξ and η. K depends not only on x, but also on this two-dimensional direction.

There is another scalar quantity obtainable from the curvature tensor, not as important as K geometrically, but of some significance for applications, especially for the theory of relativity. We first use contraction to obtain the so-called *Ricci tensor*,

$$R_{jk} = R^i{}_{jki}, \tag{11.16}$$

and we then obtain, by inner multiplication with g^{jk}, the so-called *scalar curvature* or *curvature invariant*,

$$R - g^{jl}\,R_{jl} \tag{11.17}$$

Both the Ricci tensor and the scalar curvature can be interpreted geometrically as certain mean values of the Gaussian curvatures (11.15). Let $\xi^i_{(1)}, \ldots, \xi^i_{(n)}$ be n mutually orthogonal unit vectors at the point (x) in question and let $K(r, s)$ be the Riemannian curvature (11.15) belonging to the direction of the plane spanned by $\xi^i_{(r)}, \xi^i_{(s)}$. Then

$$K(r, s) = R_{ijkl}\,\xi^i_{(s)}\,\xi^l_{(s)}\,\xi^j_{(r)}\,\xi^k_{(r)}, \tag{11.18}$$

and, since

$$g^{il} = \sum_{s=1}^{n} \xi^i_{(s)}\,\xi^l_{(s)}$$

holds, it follows from (11.16) that

$$\sum_{s=1}^{n} K(1, s) = R_{jk}\,\xi^j_{(1)}\,\xi^k_{(1)}.$$

Since $K(1, 1) = 0$, the quantity

$$\frac{1}{n-1}\,R_{jk}\,\xi^j_{(1)}\,\xi^k_{(1)}$$

is the arithmetic mean of the curvatures of $n - 1$ mutually orthogonal planes through $\xi^i_{(1)}$, and it is remarkable that this mean value is independent of the choice of the vectors $\xi^i_{(2)}, \ldots, \xi^i_{(n)}$. If we also add over all r in (11.18), it follows further that

$$R = \sum_{r,s=1}^{n} K(r, s), \tag{11.19}$$

which means that the scalar curvature R, divided by $n(n - 1)$, is the arithmetic mean of all the Riemannian curvatures belonging to the planar directions that are obtainable from n mutually orthogonal (vector) directions.

Let us assume the particular case that at a point (x), all the Riemannian curvatures belonging to all possible planar directions are equal, so that K depends only on (x) and not on the planar direction at (x). In this case,

$$K(x) = \frac{R}{n(n-1)},$$

and this formula applies to the general case if $n = 2$, since, for $n = 2$, there is only one planar direction at (x) altogether. Thus the last formula yields, for the case of the theory of surfaces,

$$K(x) = \frac{R}{2} = \frac{1}{2} g^{jk} g^{il} R_{ijkl}.$$

The reader can verify this independently, using the previously obtained expressions for the Gaussian curvature K of a surface and using the symmetry properties of the curvature tensor.

Exercise 11.6.1. *Prove* that $R_{ijkl} + R_{jikl} = 0$. (*Hint.* Prove this in special coordinates; in geodesic coordinates, i.e., if $\Gamma = 0$, we have

$$2R_{ijkl} = g_{ik,j,l} + g_{jl,i,k} - g_{il,j,k} - g_{jk,i,l}.)$$

Exercise 11.6.2. *Prove* that K is indeed dependent only on x and on the direction of the two-dimensional surface at x. (*Hint.* Replace ξ^i and η^i by $\lambda\xi^i + \mu\eta^i$ and $\nu\xi^i + \rho\eta^i$, respectively, and make use of the symmetry properties of the curvature tensor, in particular, also of Exercise 11.6.1.)

11.7 Formal Completion of the Local Theory of Surfaces in Euclidean 3-Space

From the vantage point we have now gained, we can look back over the theory of surfaces and round it off in some important aspects. First, we shall use covariant differentiation to write the equations of Gauss and Weingarten, and the equations of Gauss and Mainardi-Codazzi, in a very convenient and concise form. We shall then prove that the two tensors g_{ik} and L_{ik} determine a surface uniquely to within rigid motions.

First, the equations for the derivatives can be written as follows:

$$\begin{aligned} \mathfrak{x}_{i;k} &= L_{ik}\,\mathfrak{N} && \text{(Gauss)},\\ \mathfrak{N}_i &= -\,L_i^r\,\mathfrak{x}_r && \text{(Weingarten)}. \end{aligned} \tag{11.20}$$

The integrability conditions for this system of partial differential equations are obtained (see Appendix II.2) from the requirement that the second mixed partial derivatives of $\mathfrak{N}$ and $\mathfrak{x}_i$ should be independent of the order of the differentiations. We shall here carry this through in a different way from before (§5.4), by using covariant differentiation and; in particular, the formula for the difference of the second mixed covariant derivatives of §9.6. According to this, we have, on the one hand,

$$\mathfrak{x}_{i;k;j} - \mathfrak{x}_{i;j;k} = -\,R^m{}_{ikj}\,\mathfrak{x}_m,$$

which, on the other hand, on account of (11.20), is

$$\begin{aligned} &= (L_{ik;j} - L_{ij;k})\,\mathfrak{N} + L_{ik}\,\mathfrak{N}_j - L_{ij}\,\mathfrak{N}_k\\ &= (L_{ik;j} - L_{ij;k})\,\mathfrak{N} - (L_{ik}\,L_j^m - L_{ij}\,L_k^m)\,\mathfrak{x}_m, \end{aligned}$$

whence, equating corresponding coefficients,

$$R_{hikj} = L_{ik}\,L_{hj} - L_{ij}\,L_{hk} \tag{11.21}$$

$$L_{ik;j} = L_{ij;k}. \tag{11.22}$$

Only one of the relations (11.21) is essential, on account of the symmetry properties, and that one is, of course, just the equation of Gauss that demonstrates the *theorema egregium* and that can be written, say,

$$R_{2112} = L_{11}\,L_{22} - L^2{}_{12}.$$

The relations (11.22) of Mainardi and Codazzi contain only two essentially distinct statements, namely, for $i = k = 1, j = 2$, and for $i = k = 2, j = 1$. Thus, there are altogether three integrability conditions. To be sure, we have not taken into account here the integrability conditions that follow from Weingarten's equations; but these conditions turn out to be the same as the equations of Mainardi and Codazzi:

$$\begin{aligned} 0 = \mathfrak{N}_{i;k} - \mathfrak{N}_{k;i} &= -\,(L_i^r\,\mathfrak{x}_{r;k} - L_k^r\,\mathfrak{x}_{r;i}) - (L_{i;k}^r - L_{k;i}^r)\,\mathfrak{x}_r\\ &= -\,(L_i^r\,L_{rk} - L_k^r\,L_{ri})\,\mathfrak{N} - g^{rj}(L_{ij;k} - L_{kj;i})\,\mathfrak{x}_r. \end{aligned}$$

Here the coefficients of $\mathfrak{N}$ are zero. In obtaining the coefficients of $\mathfrak{x}_r$, Ricci's lemma was used as the last step; equating these coefficients to zero yields, once more, Eqs. (11.22).

Finally, we shall prove the theorem of O. Bonnet (1867) to the effect that the g_{ik} and L_{ik}, connected as they are by the three equations of (11.21) and (11.22), constitute a complete system of invariants for the surface:

Fundamental Theorem of the Theory of Surfaces (O. Bonnet, 1867). *Let there be assigned two symmetric tensors, $\tilde{g}_{ik}(u^1, u^2)$ of class C^2 and $\tilde{L}_{ik}$ of class C^1, and assume that they satisfy relations* (11.21) *and* (11.22) *and, furthermore, that $\tilde{g}_{ik}\,\xi^i\,\xi^k > 0$ for $\xi^i \neq 0$. Then there is one and, to within rigid motions, only one surface $\mathfrak{x}(u^1, u^2)$ of class C^3 whose fundamental forms are, in a suitable coordinate system,*

$$g_{ik} = \tilde{g}_{ik}, \qquad L_{ik} = \tilde{L}_{ik}.$$

Proof. To begin with, the system of equations for the derivatives

$$\begin{aligned} \tilde{\mathfrak{x}}_{i;k} &= \tilde{L}_{ik}\,\tilde{\mathfrak{N}} \\ \tilde{\mathfrak{N}}_i &= -\tilde{L}_i^r\,\tilde{\mathfrak{x}}_r \end{aligned} \tag{11.23}$$

has one and only one system of solutions $\tilde{\mathfrak{x}}_i$, $\tilde{\mathfrak{N}}$ belonging to assigned initial conditions

$$\tilde{\mathfrak{x}}_i(\overset{0}{u}) = \overset{0}{\mathfrak{x}}_i, \qquad \tilde{\mathfrak{N}}(\overset{0}{u}) = \overset{0}{\mathfrak{N}}$$

that satisfy $\overset{0}{\mathfrak{x}}_i\overset{0}{\mathfrak{x}}_k = \tilde{g}_{ik}(\overset{0}{u})$, $\overset{0}{\mathfrak{N}}\overset{0}{\mathfrak{x}}_i = 0$, $\overset{0}{\mathfrak{N}}{}^2 = 1$, since the integrability conditions for (11.23) are satisfied on account of the hypotheses. (For the relevant theorem on systems of partial differential equations, see Appendix II.2.) We organize the rest of the proof into several steps.

1. There exists $\mathfrak{x}(u^i)$ of class C^3 and satisfying $\partial\mathfrak{x}/\partial u^i = \tilde{\mathfrak{x}}_i$, $\mathfrak{x}(\overset{0}{u}) = \overset{0}{\mathfrak{x}}$. For, we have $\partial\tilde{\mathfrak{x}}_i/\partial u^k - \partial\tilde{\mathfrak{x}}_k/\partial u^i = \tilde{\mathfrak{x}}_{i;k} - \tilde{\mathfrak{x}}_{k;i} = 0$ on account of the symmetry of the $\tilde{L}_{ik}$, while the fact that the solution $\mathfrak{x}(u^i)$ is of class C^3 follows from (11.23) and from the assumptions about $\tilde{g}_{ik}$ and $\tilde{L}_{ik}$.

This step already shows that, if the initial point and the vectors $\overset{0}{\mathfrak{x}}$, $\overset{0}{\mathfrak{x}}_i$, $\overset{0}{\mathfrak{N}}$ at this point are assigned, there is at most one surface of the required kind, namely the $\mathfrak{x}(u^i)$ that was obtained. (Quantities without $\sim$ always refer to this surface.)

2. We have $g_{ik} = \tilde{g}_{ik}$, $\mathfrak{N} = \tilde{\mathfrak{N}}$.

For, to begin with, it follows from (11.23) that

$$(\mathfrak{x}_i\,\mathfrak{x}_j)_{;k} = \tilde{L}_{ik}(\tilde{\mathfrak{N}}\mathfrak{x}_j) + \tilde{L}_{jk}(\tilde{\mathfrak{N}}\mathfrak{x}_i)$$

$$(\tilde{\mathfrak{N}}\mathfrak{x}_j)_{;k} = -\,\tilde{L}^r_k(\mathfrak{x}_r\,\mathfrak{x}_j) + \tilde{L}_{jk}\,(\tilde{\mathfrak{N}}^2)$$

$$(\tilde{\mathfrak{N}}^2)_{;k} = -\,2\,\tilde{L}^r_k(\tilde{\mathfrak{N}}\mathfrak{x}_r).$$

Writing φ_{ij} for $\mathfrak{x}_i\,\mathfrak{x}_j$, also φ_j for $\tilde{\mathfrak{N}}\mathfrak{x}_j$ and φ for $\tilde{\mathfrak{N}}^2$, we see that we have here a system of partial differential equations for the quantities just introduced, a system once again which has at most one system of solutions corresponding to a given set of initial conditions. Now, by their construction, the quantities $\varphi_{ij} = \mathfrak{x}_i\,\mathfrak{x}_j$, $\varphi_j = \tilde{\mathfrak{N}}\mathfrak{x}_j$, $\varphi = \tilde{\mathfrak{N}}^2$ are solutions that correspond to the following initial conditions at $u = \overset{0}{u}$:

$$\varphi_{ij}(\overset{0}{u}) = g_{ij}(\overset{0}{u}), \qquad \varphi_j(\overset{0}{u}) = 0, \qquad \varphi(\overset{0}{u}) = 1;$$

on the other hand, using Ricci's lemma, we see that the quantities $\varphi_{ij} = \tilde{g}_{ij}$, $\varphi_j = 0$, $\varphi = 1$ are also a system of solutions belonging to the same initial conditions. Hence the two sets of quantities coincide, which establishes the statement under 2.

3. We have $L_{ik} = \tilde{L}_{ik}$. For, on the one hand, our surface satisfies $\mathfrak{x}_{i;k} = L_{ik}\,\mathfrak{N}$, by Gauss' equation (11.20); on the other hand, it follows, from the hypotheses and by substituting the results of step 2, that $\mathfrak{x}_{i;k} = \tilde{L}_{ik}\,\mathfrak{N}$.

4. The fact that $\mathfrak{x}(u^i)$ is of class C^3 follows directly from Eqs. (11.20) for the derivatives. The proof of the fundamental theorem is now completed by the observation that any initial point $\overset{0}{\mathfrak{x}}$ can be moved to any other one by a translation and that subsequently any initial trihedron can be mapped by a rotation onto any other one (with $\overset{0}{\mathfrak{x}}_i\,\overset{0}{\mathfrak{x}}_k = \overset{0}{g}_{ik}$), and that the fundamental tensors are invariant under rigid motions.

For the global theory of surfaces, the uniqueness part of the theorem is especially important. We restate this part as follows:

If two surfaces $\mathfrak{x}(u^i)$ and $\mathfrak{x}^*(u^i)$ are such that the two fundamental tensors of one surface agree with those of the other at all points (u^i), then the two surfaces can be brought into coincidence by a rigid motion of one of them, that is, the two surfaces are "congruent".

Exercise 11.7.1. Beltrami defined a differential operator $\Delta\varphi \underset{\text{def}}{=} g^{ik}\varphi_{,i;k}$ which, for the case of Euclidean spaces, reduces to the Δ-operator ($\Delta = \Sigma\partial^2/\partial x_k^2$). *Show* that for surfaces,

(a) $\Delta\mathfrak{x} = 2H\mathfrak{N}$ (hence, $\Delta\mathfrak{x} = 0$ only for minimal surfaces!);

(b) $\Delta\mathfrak{N} = -2H_{,j}\,g^{jk}\,\mathfrak{x}_k - 2(2H^2 - K)\mathfrak{N}$.

Show also that $\mathfrak{N}\Delta\mathfrak{N} \leqq 0$, with equality holding only for the plane.

[*Hint.* Use the formula $H = \frac{1}{2}g^{ik}L_{ik}$, also Eqs. (11.20) for the derivatives, and the equations of Codazzi. For (b), it is most convenient to use the lines of curvature as the parametric net, so that $L_{12} = g_{12} = 0$.]

Remarks on Complete Systems of Invariants. By the fundamental theorem, is is true that the g_{ik} and L_{ik} determine a surface completely; but these quantities are neither independent (since they are related by the equations of Gauss and of Codazzi), nor are they scalar invariants. Indeed, the problem of finding a complete system of independent scalar invariants is much more difficult than the corresponding problem in the theory of curves. For recent investigations (also in higher dimensions) and for references to older literature, the interested reader may consult papers by K. Leichtweiss [*Math. Z.* **57**, 244 (1953); *Math. Ann.* **132**, 1 (1956)]. Assuming analyticity, Leichtweiss proved that a hypersurface in a Riemannian space is determined by assigning a linear fractional function of its principal radii in geodesic parallel coordinates, and an initial strip. Related results are by F. Rellich [*Math. Z.* **43**, 618 (1938): H as a function of suitably normalized isothermal parameters and K as a function of suitably normalized asymptotic-curve parameters determine a surface]; and by W. Süss [*Arch. Math.* **2**, 103 (1949): The sum of the radii of principal curvature and the third fundamental form — i.e., $\gamma_{ik} = \mathfrak{N}_i\,\mathfrak{N}_k$ — determine a surface; classical results in this direction are by E. B. Christoffel].

11.8 The Geodesic Deviation

Let $x^i(s;\,\varepsilon)$ be a family of geodesics, with initial conditions $x^i(0;\,\varepsilon)$ and $x^{i\prime}(0;\,\varepsilon) = \xi^i(0;\,\varepsilon)$. Let us assume that the metric is of class C^ω and that the initial conditions are analytic in the parameter ε. Then the $x^i(s;\,\varepsilon)$ are solutions of the system of analytic differential equations

$$x^{i\prime\prime} + \Gamma^i_{jk}\,x^{j\prime}\,x^{k\prime} = 0 \tag{11.24}$$

and are therefore analytic in ε, so that, for $|\varepsilon|$ sufficiently small,

$$x^i(s;\,\varepsilon) = x^i(s) + \varepsilon\,v^i(s) + \varepsilon^2(\ldots) \tag{11.25}$$

$$\Gamma^i_{jk}(x(s;\,\varepsilon)) = \Gamma^i_{jk}(x(s)) + \Gamma^i_{jk,l}(x)\,\varepsilon\,v^l(s) + \varepsilon^2(\ldots). \tag{11.26}$$

If we substitute this into the differential equations (11.24), we obtain [with all functions to be taken at $x(s)$]

$$v^{i\prime\prime}(s) + 2\,\Gamma^i_{jk}\,v^{j\prime}\,x^{k\prime} + \Gamma^i_{jk,l}\,x^{j\prime}\,x^{k\prime}\,v^l = \varepsilon(\ldots). \tag{11.27}$$

Let us write the left-hand side of (11.27) in covariant notation. Since

$$\frac{Dv^i}{ds} = v^{i\prime} + \Gamma^i_{jk}\, v^j\, x^{k\prime} \qquad \text{and} \qquad \frac{D^2v^i}{ds^2} = v^{i\prime\prime} + \Gamma^i_{jk,l}\, x^{l\prime}\, x^k\, v^j$$

$$+ 2\,\Gamma^i_{jk}\, v^{j\prime}\, x^{k\prime} + \Gamma^i_{jk}\, v^j\, x^{k\prime\prime} + \Gamma^i_{lm}\, \Gamma^l_{jk}\, v^j\, x^{k\prime}\, x^{m\prime}$$

hold, and since we can use $x^{k\prime\prime} = -\,\Gamma^k_{lm}\, x^{l\prime}\, x^{m\prime}$, we obtain in place of (11.27)

$$\frac{D^2v^i}{ds^2} + R^i{}_{jkl}\, v^l\, x^{j\prime}\, x^{k\prime} = \varepsilon(\ldots). \tag{11.28}$$

If we let ε go to 0 in (11.28), we obtain a differential equation that is known as the equation of the geodesic deviation [T. Levi-Cività, *Math. Ann.* **97**, 291–320 (1927)]; it governs the deviation from a given geodesic of neighboring geodesics.

For the two-dimensional case, this equation goes back to Jacobi. In this case, let us consider, in particular, geodesics that emanate from a fixed point, so that $x^i(0;\varepsilon) = x^i(0)$. We can then show, first, that v^i is orthogonal to the geodesic $x^i(s)$. In fact, using the differential equation of geodesics, we find that

$$(g_{ik}\, v^i\, x^{k\prime})' - g_{ik}\, v^{i\prime}\, x^{k\prime} + \frac{1}{2}\frac{\partial g_{rj}}{\partial x^i}\, v^i\, x^{r\prime}\, x^{j\prime}, \tag{11.29}$$

and, since s is the arc length on all curves, it follows that

$$1 = g_{ik}(x(s;\varepsilon))\, x^{i\prime}(s;\varepsilon)\, x^{k\prime}(s;\varepsilon)$$

$$= 1 + \varepsilon\left(2\, g_{ik}\, v^{i\prime}\, x^{k\prime} + \frac{\partial g_{ik}}{\partial x^l}\, v^l\, x^{i\prime}\, x^{k\prime}\right) + \varepsilon^2(\ldots),$$

whence it follows that the right-hand side of (11.29) vanishes. Therefore, from (11.29),

$$g_{ik}\, v^i\, x^{k\prime} = 0,$$

because this holds true at $s = 0$.

Hence, if n^i is the unit normal vector taken along the curve $x^i(s)$, it follows that

$$v^i(s) = v(s)\, n^i(s).$$

Let us substitute this into the differential equation of the geodesic deviation [i.e., (11.28) for $\varepsilon \to 0$], making use of the fact that the covariant derivative of n^i vanishes. (This is so because the tangent vectors are geodesically parallel along the geodesic; therefore, so are the unit vectors perpendicular to them.) It follows that

$$\frac{d^2v}{ds^2} n^i + v \cdot R^i{}_{jkl} n^l x^{j'} x^{k'} = 0.$$

If the coordinate system is chosen in such a way that the initial geodesic is the base curve C of a geodesic coordinate system (see §4.5), then we have

$$x^{2'} = n^1 = 0, \qquad x^{1'} = n^2 = 1$$

and we therefore obtain

$$v'' + K(s)\, v = 0 \tag{11.30}$$

as the differential equation of Jacobi. The behavior of the solutions of this differential equation is studied in detail in the theory of differential equations (see for example E. Kamke, "Differentialgleichungen," 5th ed., Vol. 1, p. 128 ff. Leipzig, 1956, or G. Birkhoff and G.-C. Rota, "Ordinary Differential Equations," p. 247 ff. Boston, 1962). In particular, if $K(s) \geqq q > 0$ holds along the initial geodesic, then geodesics with neighboring initial conditions will not deviate far from it, but will intersect it again and again. The study of the geodesic deviation is of interest for stability problems in mechanics; we shall return to this later (in §14.5).

CHAPTER IV

Further Development and Applications of Riemannian Geometry

Up to this point, we were interested in Riemannian geometry largely for the sake of constructing a framework within which a consistent intrinsic geometric development of the geometry of a surface could be carried out. However, Riemannian spaces are of geometric interest beyond this and we now turn to studying them in their own right. Of particular interest are the spaces of constant curvature, to which §12 is devoted. They are intimately connected with classical non-Euclidean geometry, which is historically the first example of a geometric structure other than Euclidean geometry. This non-Euclidean geometry was developed in the first half of the nineteenth century by Bolyai, Lobachevski, and Gauss, and differs from Euclid's geometry by having a different axiom of parallels: Through a point P of the plane not on a line g, there is more than one line that does not intersect g (i.e., there are at least two "parallels" to g through P). This non-Euclidean geometry is a special Riemannian geometry. Thus Riemann's achievements in geometry are not confined to giving a satisfactory foundation for Gauss' intrinsic geometry of surfaces but include also a contribution to the deeper understanding of non-Euclidean geometry.

The reader should always bear in mind that a chief aim of any generalization in geometry is a more profound appreciation of the old and well known. Thus in §15 we shall study more general differential-geometric systems and we shall then be able to appreciate Riemannian geometry from a new vantage point, from which we can perceive some of its distinguishing features.

As to another matter, namely the question of applications of geometry to other disciplines, we cannot give it more than very limited space in this book. However, in §14 we want to get at least a glimpse of the diversity of applications of Riemannian geometry to analytical mechanics. Before this, in §13, we shall have to study mappings of Riemannian spaces onto each other. In this study, some important theorems of the theory of surfaces will find their natural setting.

§ 12. Spaces of Constant Curvature and Non-Euclidean Geometry

12.1 Spaces of Constant Curvature

A Riemannian space is said to be of constant curvature if the value of the Riemannian curvature K (see §11.6) depends neither on the point (x) nor on the planar direction at (x). The first fact we shall prove is that, for any real number K and any integer $n \geqq 2$, there exists a Riemannian space of dimension n and of constant curvature K. For $n = 2$, we know this to be true from the theory of surfaces (§6.4). Now let n be $\geqq 3$ and take K positive, to begin with; we set $K = 1/\rho^2$. We shall show that the sphere S_n of radius ρ in Euclidean $(n + 1)$-dimensional space is a Riemannian space of constant curvature $K = 1/\rho^2$.

This can be shown without any computation. We first establish the fact that any plane E_2 through the center of the sphere intersects S_n in a curve C which is a geodesic of S_n. To this end, consider the reflection of the space in the plane E_2 (that is, the affine transformation that leaves E_2 pointwise invariant and maps every vector $\mathfrak{v}$ that is orthogonal to E_2 onto the vector $-\mathfrak{v}$). This reflection leaves C invariant point by point. Also, being an isometric mapping of S_n onto itself, it maps geodesics onto geodesics. If C' is the uniquely determined geodesic that is tangent to the circle C at a point P of C, then the image of C' under the reflection is also a geodesic tangent to C at P and must therefore coincide with C'; hence $C' = C$, because the only fixed points under the reflection are those of C itself.

Let the origin of the coordinate system be the center of the sphere, let the position vector of a point P of S_n be $\mathfrak{x}$, and let $\mathfrak{y}$, $\mathfrak{z}$ be two tangent vectors to S_n at P. If we intersect S_n with the three-dimensional Euclidean space E_3 spanned by $\mathfrak{x}$, $\mathfrak{y}$, and $\mathfrak{z}$, we obtain as the intersection an S_2 that contains all the geodesics of S_n that emanate from P in the planar direction spanned by $\mathfrak{y}$ and $\mathfrak{z}$. Hence the Riemannian curvature K of S_n at P in this planar direction is the Gaussian curvature of the above S_2, and this is indeed equal to $1/\rho^2$.

For an analytic verification, we write S_n in the form

$$x_{n+1}^2 = \rho^2 - (x_1{}^2 + x_2{}^2 + \ldots + x_n{}^2)$$

and we can use the coordinates $x_1, \ldots, x_n$ as our (independent) coordinates of the manifold S_n. For the element of arc we then have

$$\left(\frac{ds}{dt}\right)^2 = \dot{\mathfrak{x}}^2 = \dot{x}_1^2 + \ldots + \dot{x}_{n+1}^2 = g_{ij}\, \dot{x}^i\, \dot{x}^j$$

(where the summations over i, j are from 1 to n), whence

$$g_{ij} = \delta_{ij} + \frac{x_i\, x_j}{(1/K) - \delta_{hm}\, x^h\, x^m}.$$

The parametrization is admissible for, say, the "northern" hemisphere

$$\delta_{ij}\, x^i\, x^j < \rho^2.$$

As for the rest, we leave it to the reader to verify, as a computational exercise, that this metric does, indeed, give Riemannian curvature $K = \text{const.}$ It turns out that if $K < 0$, the metric tensor g_{ij} exhibited above again leads to Riemannian curvature $K = \text{const.}$, so that we obtain, at the same time, examples of Riemannian metrics with constant negative curvature. Finally, for the case $K = 0$ we can take the Euclidean spaces themselves as examples, and this completes the proof of the fact that there are Riemannian spaces of any given constant curvature.

12.2 A Theorem of Schur on Spaces of Constant Curvature

We shall now prove a result of F. Schur (1886) to the effect that K cannot be independent of the planar directions at each point x unless it is also independent of x or, stated more precisely: If $n \geqq 3$ and if K is a function of x alone, $K = K(x)$, then $K = \text{const.}$ (For $n = 2$, this does not apply; we know that there are surfaces with nonconstant Gaussian curvature.)

Toward this result, we first prove the following:

Theorem 12.2.1. *If $K = K(x)$, then*

$$R_{ijkl} = K(x)(g_{il}\, g_{jk} - g_{ik}\, g_{jl}) \tag{12.1}$$

or, equivalently,

$$R^i{}_{jkl} = K(x)(\delta^i_l\, g_{jk} - \delta^i_k\, g_{jl}). \tag{12.2}$$

Proof. Consider the tensor

$$T_{ijkl} = R_{ijkl} - K(x)(g_{il}\, g_{jk} - g_{ik}\, g_{jl}),$$

which has the same symmetry properties as R_{ijkl}, as can be verified immediately. By (11.15), we have for arbitrary ξ, η that

$$T_{ijkl}\, \eta^i\, \eta^k\, \xi^j\, \xi^l = 0.$$

In particular, if we set

$$\eta^i = \alpha\, \delta^i_r + \beta\, \delta^i_t, \qquad \xi^j = \gamma\, \delta^j_s + \varepsilon\, \delta^j_u,$$

then all of the coefficients of the resulting polynomial in $\alpha, \beta, \gamma, \varepsilon$ must vanish. Applying this, in particular, to the coefficient of the product $\alpha\beta\gamma\varepsilon$, we find that

$$T_{rstu} + T_{ruts} + T_{tsru} + T_{turs} = 0$$

or, since $T_{rstu} = T_{turs}$,

$$T_{turs} + T_{tsru} = 0, \tag{12.3}$$

and similarly $T_{trsu} + T_{tusr} = 0$. On the other hand, we can interchange the last two subscripts on each of the terms in (12.3) and obtain $T_{tusr} + T_{tsur} = 0$, hence $T_{tsur} = T_{trsu}$, and similarly $T_{trsu} = T_{turs}$ by a cyclic permutation of the subscripts. Now it follows from Bianchi's identity that $T_{tsur} = 0$, which is what the theorem asserts.

We can now prove

Theorem 12.2.2 (F. Schur). *If $K = K(x)$ and $n \geqq 3$, then $K =$* const.

Proof. By Theorem 12.2.1,

$$R^i{}_{hjk} = K(\delta^i_k\, g_{hj} - \delta^i_j\, g_{hk}),$$

whence, by Ricci's lemma and by $\delta^i_{k;l} = 0$ (see Exercise 9.5.2),

$$R^i{}_{hjk;l} = \frac{\partial K}{\partial x^l}\,(\delta^i_k\, g_{hj} - \delta^i_j\, g_{hk}).$$

We form the corresponding expressions resulting from cyclic permutations of j, k, l and then add, and using also Bianchi's identity (9.10) we then obtain

$$0 = K_{,l}(\delta^i_k\, g_{hj} - \delta^i_j\, g_{hk}) + K_{,k}(\delta^i_j\, g_{hl} - \delta^i_l\, g_{hj}) + K_{,j}(\delta^i_l\, g_{hk} - \delta^i_k\, g_{hl})$$

which, after contraction with respect to i and k and inner multiplication by g^{hj}, yields

$$0 = K_{,l}(n-1)\,(n-2).$$

This proves the theorem.

The significance of this theorem in a more general context will be discussed later (in §15.9).

12.3 Free Mobility in the Two-Dimensional Case

We consider a two-dimensional Riemannian space referred to a system of geodesic parameters u^1, u^2 which is such that the base curve C of the construction in §4.5 is itself a geodesic curve. Thus the parametric net consists of the geodesic C plus the geodesics orthogonal to C plus the curves that are geodesically parallel to C. If C is given by $u^1 = 0$ and if we choose u^2 as the arc length along C, then

$$g_{11} = 1, \qquad g_{12} = 0, \qquad g_{22} = f^2(u^i), \qquad f(0, u^2) = 1, \qquad \frac{\partial f(0, u^2)}{\partial u^1} = 0. \quad (12.4)$$

In these geodesic parameters, the Gaussian curvature is given (see Exercise 5.5.5) by

$$K = -\frac{1}{f}\frac{\partial^2 f}{\partial (u^1)^2} \tag{12.5}$$

For an assigned value of K and for any fixed u^2, this is an ordinary differential equation for f whose solution is uniquely determined by K and by the initial conditions (12.4). We have therefore proved the following result:

Theorem 12.3.1. *If two two-dimensional spaces have geodesic parametrizations* (12.4) *such that the base curves C and $\bar{C}$ are geodesics, then a necessary and sufficient condition for (sufficiently small pieces of) the two spaces to be isometric is that their Gaussian curvatures should be the same at corresponding points.*

Stated differently, this says that the Gaussian curvature determines the metric uniquely in the particular type of geodesic parametrization that was used. Note, however, that the proviso concerning the special parametrization is quite essential; it is not true that every mapping which preserves the curvature at corresponding points is an isometry. (Example: An arbitrary mapping of the plane onto itself.) If we apply Theorem 12.3.1 to the special case of spaces of constant curvature, we obtain the following result:

Theorem 12.3.2. *Let V_2 and $\bar{V}_2$ be two two-dimensional spaces of the same constant curvature K and let ξ (and $\bar{\xi}$, respectively) be a direction at a point P (and $\bar{P}$, respectively) of V_2 (and $\bar{V}_2$, respectively). Then there exists an isometric mapping from a neighborhood of P to a neighborhood of $\bar{P}$ under which P goes into $\bar{P}$ and ξ into $\bar{\xi}$.*

To prove this, it suffices to choose systems of geodesic parallel coordinates based on the two geodesic curves in the given directions.

If the two spaces are the same, then the last result implies the free mobility in spaces of constant curvature: If P and $\bar{P}$ are any two points of a space V_2 of constant curvature and if ξ and $\bar{\xi}$ are any given directions at P and $\bar{P}$, respectively, then, for any (sufficiently small) neighborhood of P, there is a congruent neighborhood of $\bar{P}$ such that ξ and $\bar{\xi}$ correspond to each other under the congruence. This state of affairs for spaces of constant curvature gives them an aspect shared by Euclidean plane geometry. And it is in fact true that from the result on free mobility one can derive the axioms of congruence of Euclidean geometry provided that the role of Euclidean straight lines is assigned to geodesics.[1] We shall carry through the derivation of the "side-angle-side" congruence theorem and we shall leave it to the reader to derive the remaining congruence axioms (say, those of Hilbert's system, in his "Grundlagen der Geometrie," 8th ed., Stuttgart, 1956).

Theorem 12.3.3. *In a V_2 of constant curvature, let P, Q, R and P_1, Q_1, R_1 be two triangles with geodesics as sides and lying in sufficiently small neighborhoods of P and P_1, respectively. Assume that the geodesic distances satisfy $PQ = P_1Q_1$, $PR = P_1R_1$ and, furthermore, that the angles at P and P_1 are the same. Then $QR = Q_1R_1$ must also hold and, furthermore, corresponding angles, at the two remaining vertices, are equal.*

The proof follows from the existence of an isometric mapping that maps P onto P_1 and maps the geodesic PQ onto the geodesic P_1Q_1, and which can also be selected in such a way that it maps R onto R_1 (a "reflection").

Of the Euclidean axioms, the only one that is not satisfied if $K \neq 0$ is the axiom of parallels. Specifically, if $K > 0$ then through a point not on a given geodesic there does not exist any geodesic that fails to intersect the given geodesic (as is easily verified for a sphere). But if $K < 0$ then there exist infinitely many, as we shall show shortly, after first studying a few other items such as, in particular, the generalization of the results on free mobility to the case of n dimensions.

These results also include the characterization of the spaces of constant curvature by the property of free mobility, as follows:

---

[1] In this, one has to assume *completeness* of the space, i.e., the requirement that every geodesic curve can be extended indefinitely. The sphere is an example of a complete space with $K > 0$. The pseudosphere (§6.4), on the other hand, satisfies $K = \text{const.} < 0$, but is not complete. We shall see in §12.5 that for any constant K, there exist complete spaces.

Theorem 12.3.4. *For a space V_2 to be of constant curvature, it is necessary and sufficient that, for any two of its points, there exist suitable neighborhoods that can be mapped isometrically onto each other.*

For if such isometries exist for any pair of points of the space, then, by the theorema egregium, the Gaussian curvature is the same at any point as at any other point.

Exercise 12.3.5. *Prove* that if $K = \text{const.}$, then the differential equation (12.5) with initial conditions (12.4) can be solved explicitly, its solution being

(a) $f = \cos \sqrt{K}\, u^1$ if $K > 0$,

(b) $f = 1$ if $K = 0$,

(c) $f = \cosh \sqrt{|K|}\, u^1$ if $K < 0$.

Exercise 12.3.6. From Exercise 12.3.5, case (c), it follows that a complete space V_2 exists for any given constant negative curvature. In fact, the parameters u^i are then admissible for all values between $-\infty$ and $+\infty$, because $g > 0$.

12.4 The Isometry of Spaces of Constant Curvature

We shall now extend the results of the preceding subsection to any number of dimensions. We could again make use of some special coordinate system, say, of normal coordinates, but we shall find it more convenient to prove the following result in general coordinates.

Theorem 12.4.1. *Given the following: Two Riemannian spaces V_n and $\bar{V}_n$ of the same constant curvature, a point P in V_n and a point $\bar{P}$ in $\bar{V}_n$, an orthonormal n-tuple $\xi^i_{(1)}, \ldots, \xi^i_{(n)}$ of vectors at P and another, $\bar{\xi}^i_{(1)}, \ldots, \bar{\xi}^i_{(n)}$, at $\bar{P}$. Then there is an isometric mapping $\bar{x}^i = \bar{x}^i(x)$ of a neighborhood of P onto a neighborhood of $\bar{P}$ such that P and $\xi^i_{(r)}$ are mapped, respectively, onto $\bar{P}$ and $\bar{\xi}^i_{(r)}$.*

Proof. We are to find a mapping for which

$$\bar{g}_{ik} = g_{jl} A^j_i A^l_k \tag{12.6}$$

with

$$\frac{\partial x^j}{\partial \bar{x}^i} = A^j_i. \tag{12.7}$$

We can assume here (since we can always achieve this by a linear coordinate transformation in each space) that P and $\bar{P}$ are the coordinate origins and that $\xi^i_{(r)} = \bar{\xi}^i_{(r)} = \delta^i_r$, so that the initial conditions are

$$\bar{g}_{ik}(0) = g_{ik}(0) = \delta_{ik}, \qquad A^j_i(0) = \delta^i_j.$$

The integrability conditions for (12.6) cannot just be read off immediately, because these differential equations are not exhibited in a form solved explicitly for the derivatives. We first differentiate and we obtain, after some reshuffling familiar to us from the derivation of the law of transformation for the Christoffel symbols,

$$A^j_{i,r} = \bar{\Gamma}^\alpha_{ir} A^j_\alpha - \Gamma^j_{\beta\gamma} A^\beta_i A^\gamma_r. \tag{12.8}$$

From the way this was obtained, we know that every solution of (12.8) satisfies

$$\bar{g}_{ik} - g_{jl} A^j_i A^l_k = c_{ik} = \text{const.},$$

and, since our initial conditions imply that $c_{ik}(0) = 0$, it follows that $c_{ik} \equiv 0$ so that (12.8) together with (12.7) is equivalent to (12.6) together with (12.7). The integrability conditions for (12.7) are satisfied on account of the symmetry of the right-hand side of (12.8). The integrability conditions for (12.8) are

$$\bar{R}_{ijkl} = R_{\alpha\beta\gamma\delta} A^\alpha_i A^\beta_j A^\gamma_k A^\delta_l. \tag{12.9}$$

Here we may substitute (by the general theory, see Appendix II.2) from the differential equations themselves and thus, in the present case, also from the equivalent equations (12.6). Since by hypothesis

$$\bar{R}_{ijkl} = K(\bar{g}_{il}\,\bar{g}_{jk} - \bar{g}_{ik}\,\bar{g}_{jl})$$

and

$$R_{\alpha\beta\gamma\delta} = K(g_{\alpha\delta}\,g_{\beta\gamma} - g_{\alpha\gamma}\,g_{\beta\delta}),$$

substitution into (12.9) yields conditions that hold identically, on account of (12.6), and this proves the integrability of the system.

It is an important consequence of the theorem just proved that all Riemannian spaces of constant curvature are locally isometric to the spherical spaces considered in §12.1. This fact can be used for simple proofs of many theorems on spaces of constant curvature.

If V_n and $\bar{V}_n$ are one and the same space, we can apply the last theorem to prove the following result:

Theorem 12.4.2. *The spaces of constant curvature are characterized by the necessary and sufficient property of free mobility, that is, by the following property: For any two points P, P' and two orthonormal n-tuples of vectors* *at P and P', respectively, there exists a neighborhood of P that is mapped isometrically onto a neighborhood of P' in such a way that P is mapped onto P', while the assigned n-tuple at P is mapped onto that at P'.*

Proof. That the property is necessary for constancy of curvature follows from Theorem 12.4.1. That the property is also sufficient is easily seen from the definition of the Riemannian curvature K as the Gaussian curvature of a (two-dimensional) surface of geodesics. For, the geodesic surfaces spanned at P and at P' by $\xi^i_{(1)}$, $\xi^i_{(2)}$ and by $\bar{\xi}^i_{(1)}$, $\bar{\xi}^i_{(2)}$, respectively, have the same Gaussian curvature, on account of the assumed isometry and, since the points and the two-dimensional directions at them can be chosen at will, it follows that the Riemannian curvature is constant.

These results go back, in part, even to *Riemann's* lecture of 1854. After its posthumous publication in 1868, *H. von Helmholtz* took up the problems in question anew.[1] For him, as a physicist, the problem was of greatest interest in the following formulation: Our measuring is done by moving and applying standardized scales (yardsticks) in space; such yardsticks are bodies that extend in three dimensions and, for this kind of measuring process to make sense at all, the measuring bodies must be "rigid" under any of their motions, that is, all lengths in such a body must remain invariant as the body is moved around in space. By what has just been proven, such free mobility of rigid bodies is possible only in spaces of constant Riemannian curvature. Small wonder that Helmholtz' philosophy of science led him to the conclusion that the very manner in which we measure makes the metric of our space such that it is of constant curvature. Nowadays we can ascribe little more than historic significance to this formulation of Helmholtz, considering the thoroughgoing critical reappraisal of the concept of rigid body in twentieth-century physics. And, in fact, the metric of our universe — of the four-dimensional manifold of space and time — is a Riemannian metric, as used in Einstein's general theory of relativity, and this metric is by no means subject to the assumption of constant curvature.

Incidentally, Helmholtz tried without success to characterize Euclidean spaces among all spaces of constant Riemannian curvature by means of some simple geometric condition. We shall return later to the problem of finding such conditions and we only anticipate here, by mention, the result that the Euclidean spaces are the only ones that admit similarity transformations other than rigid motions.

In §15.3 we shall study the question that was Helmholtz' main concern in these matters, namely the characterization of Riemannian spaces among more general metric spaces.

---

[1] B. Riemann, "Über die Hypothesen, welche der Geometrie zugrunde liegen." New printing, with an Appendix by H. Weyl, 3rd ed., Berlin, 1923.

H. v. Helmholtz, Über die Tatsachen, welche der Geometrie zugrunde liegen. *Nachr. Ges. Wiss., Göttingen* (1868), also in Helmholtz's "Wissenschaftliche Abhandlungen" ("Scientific Works"), Vol. 2, p. 618, Leipzig, 1883.

12.5 Special Forms of the Fundamental Tensor

For later applications, we shall now investigate certain special forms of the fundamental metric tensor in spaces of constant curvature. As a by-product, we shall obtain the existence of complete spaces of negative constant curvature.

Let us try a fundamental tensor in the particular form (so-called *isothermal* form)

$$g_{ik} = \frac{\delta_{ik}}{F^2}, \tag{12.10}$$

where $F \neq 0$ is a function of x^i. It follows that $g^{ik} = F^2 \, \delta^{ik}$. In the following sets of formulas, by way of an exception in notation, no summation is to be understood and we are not allowed to substitute equal values for different indices. Using

$$\frac{\partial g_{ik}}{\partial x^j} = -\delta_{ik} \frac{2}{F^3} \frac{\partial F}{\partial x^j} = -\frac{2\delta_{ik}}{F^2} \frac{\partial \log F}{\partial x^j}$$

and setting $\log F = f$, we find, for the Christoffel symbols, that

$$\Gamma^i_{jk} = 0, \qquad \Gamma^i_{kk} = f_{,i}, \qquad \Gamma^i_{ik} = -f_{,k}, \qquad \Gamma^i_{ii} = -f_{,i} \tag{12.11}$$

and

$$\begin{aligned} &R^i{}_{jkl} = R^i{}_{ikl} = 0, \qquad R^i{}_{jil} = -f_{,j,l} - f_{,j} f_{,l} \\ &R^i{}_{jij} = -f_{,j,j} - f_{,i,i} + \sum\nolimits^* f^2_{,r}, \qquad R^i{}_{iil} = 0 \\ &R^i{}_{jjl} = f_{,i,l} + f_{,i} f_{,l}, \end{aligned} \tag{12.12}$$

where Σ^* indicates summation over all $r \neq i, \neq j$.

If the metric is to yield a constant Riemannian curvature K, then from

$$R^i{}_{jkl} = K(\delta^i_l \, g_{jk} - \delta^i_k \, g_{jl}) = \frac{K}{F^2} (\delta^i_l \, \delta_{jk} - \delta^i_k \, \delta_{jl})$$

we obtain the following necessary and sufficient conditions:

$$\begin{aligned} &f_{,j,l} + f_{,j,} f_{,l} = 0 \\ &f_{,j,j} + f_{,l,l} - \sum\nolimits^* f^2_{,r} = \frac{K}{F^2} \end{aligned} \tag{12.13}$$

or, since $f_{,j,l} + f_{,j} f_{,l} = (1/F)F_{,j,l}$ (also for $j = l$),

$$F_{,j,l} = 0 \tag{12.14}$$

and

$$F(F_{,j,j} + F_{,l,l}) = K + \sum_{r=1}^{n} F^2_{,r}. \tag{12.15}$$

We have thus obtained

Theorem 12.5.1. *The differential equations* (12.14) *and* (12.15) *constitute necessary and sufficient conditions for the metric* (12.10) *to yield constant Riemannian curvature K.*

Now (12.14) implies that $F = G_1 + \ldots + G_n$ where G_j is a function of x^j only. If we substitute this into (12.15) and into the corresponding equation with i instead of j, we obtain

$$G_i'' = G_j'',$$

where the primes denote differentiations with respect to the argument that is present. On the left, we have a function of x^i alone, on the right a function of x^j alone, hence the two sides are both equal to one and the same constant $2c$, and we have

$$G_i = c\, x_i^2 + b_i\, x_i + a_i, \qquad \text{(no summation).} \tag{12.16}$$

Thus the G_i are quadratic polynomials whose coefficients are related, on account of (12.15), by

$$K = \sum_{r=1}^{n} (4\, a_r\, c - b_r^2), \tag{12.17}$$

and we have obtained the following result:

Theorem 12.5.2. *A necessary and sufficient condition for the metric* (12.10) *to be of constant Riemannian curvature K is that* $F = G_1 + \ldots + G_n$ *where the* G_i *satisfy the relations* (12.16) *and* (12.17).

With the special choices $c = K/4$, $b_r = 0$, $a_r = 1/n$ in (12.16), we obtain the following particular metric for a space of constant curvature K — a form of the metric due to Riemann himself:

$$g_{ik} = \frac{\delta_{ik}}{[1 + (K/4)(\delta_{jl}\, x^j x^l)]^2}\,. \tag{12.18}$$

The important example of a "non-Euclidean metric for the upper half-space" is obtained, for $K = -1$, with the choices $c = a_i = 0$ and $b_n = 1$, all other $b_i = 0$:

$$g_{ik} = \frac{\delta_{ik}}{x_n^2}\,. \tag{12.19}$$

This metric makes the entire half-space $x_n > 0$ into a Riemannian space of constant negative curvature $K = -1$. We proceed to study this example in more detail in the next subsection.

12.6 Poincaré's Model of Hyperbolic Non-Euclidean Geometry

Following Poincaré (around 1882), let us investigate more closely the metric with constant curvature $K = -1$ given in $x^n > 0$ by formula (12.19).

Theorem 12.6.1. *The geodesics associated with the metric* (12.19) *are the circles whose planes are orthogonal to the hyperplane* $x^n = 0$ *and whose centers lie in that hyperplane.*

Proof. It suffices to consider circles in the (x^1, x^n)-plane, since all others in question can be transformed into this kind by an orthogonal transformation, preserving the metric, in $x^1, \ldots, x^{n-1}$. Choosing $t = x^1$ as the parameter, such a circle is given by $x^n = \sqrt{R^2 - (t-c)^2}$, and we have

$$\dot{x}^n = -(t-c)/x^n, \qquad \ddot{x}^n = -1/x^n - (t-c)^2/(x^n)^3.$$

For the Christoffel symbols we therefore have (no summation!)

$$\Gamma^n_{nn} = -1/x^n, \qquad \Gamma^n_{ii} = 1/x^n, \qquad \Gamma^i_{ni} = -1/x^n$$

with all other Christoffel symbols being zero, so that

$$\ddot{x}^i + \Gamma^i_{jk}\,\dot{x}^j\,\dot{x}^k = \mu \cdot \dot{x}^i \qquad \text{with} \qquad \mu = 2(t-c)/(x^n)^2.$$

Therefore, by Theorem 10.4.3, the circle is a geodesic. On the other hand, through every point of the half-space, there is in every direction exactly one circle of the type specified in Theorem 12.6.1 (including the straight lines orthogonal to the hyperplane $x^n = 0$); therefore, by the uniqueness theorem for geodesics, these circles comprise all of the geodesics in the given metric and the theorem is proved.

If we approach the hyperplane $x^n = 0$ along a geodesic, then the arc length will increase beyond all bounds, since

$$\int_{x_n=\varepsilon}^{x_n=C} \frac{\sqrt{(dx^1)^2 + \ldots + (dx^n)^2}}{x^n} > \int_{\varepsilon}^{C} \frac{dx^n}{x^n} = \ln C - \ln \varepsilon,$$

which tends to ∞ as $\varepsilon \to 0$. This means that one may proceed indefinitely along any geodesic and gives the following result:

Theorem 12.6.2. *The Riemannian space with the metric*

$$g_{ik} = \frac{\delta_{ik}}{(x^n)^2}, \qquad x^n > 0$$

is a complete space of constant negative curvature $K = -1$.

The above space is an adequate model of hyperbolic non-Euclidean geometry of n dimensions, a model which is especially convenient because the measurement of angles in it happens to coincide with Euclidean measurement of angles in the half-space. (This is an immediate consequence of the form of the fundamental metric tensor.)

As references for the reader who is interested in non-Euclidean geometry, we mention H. S. M. Coxeter, "Non-Euclidean Geometry," Toronto, 1947; O. Perron, "Nichteuklidische Elementargeometrie der Ebene," Stuttgart, 1962; also the little volume by Baldus-Löbell, "Nichteuklidische Geometrie," 3rd ed., Berlin, 1953 (Sammlung Göschen, No. 970). We shall on several later occasions return to this geometry, from the differential-geometric point of view. Here we only add the remark that the differential-geometric model furnishes a (relative) proof of the consistency of non-Euclidean geometry, developed axiomatically in close analogy to Euclid's structure by Gauss, Bolyai, and Lobachevski. Such consistency proofs, using different models, were also given before Poincaré by the geometers Beltrami (1868) and Felix Klein. Poincaré's model for $n = 2$ plays an important role in complex function theory.

Exercise 12.6.3. The rigid motions of a space of constant negative curvature can be represented analytically in an especially simple way for the case of Poincaré's model. In particular, if $n = 2$ we can introduce the complex variable $z = x_1 + ix_2$ and we then find that the rigid motions are given by

$$\bar{z} = (az + b)/(cz + d), \qquad a, b, c, d \text{ real}, \qquad ad - bc > 0.$$

Verify that these transformations map the upper half-plane onto itself and that they leave the metric invariant.

§ 13. Mappings

13.1 Analytic Representation of a Mapping from One Manifold to Another of the Same Dimension

A mapping from a manifold V_n to a manifold $\tilde{V}_n$ induces (and is induced by) a relation between the respective coordinates,

$$\tilde{x}^i = \tilde{x}^i(x^k); \tag{13.1}$$

here we shall always assume that (13.1) is sufficiently differentiable and that the functional determinant does not vanish, so that, locally at least, the mapping is one to one. Under these assumptions, we can use the x^i as coordinates in $\tilde{V}_n$ also. Corresponding points of the two manifolds will then have the same coordinates, and we say, in this situation, that we have a correspondence "by equal coordinates" between the two manifolds. A well-known example is furnished by geographic maps, which represent mappings from pieces of the sphere to the Euclidean plane. We usually transfer the geographic coordinates from the sphere to the image plane by drawing in, on the geographic maps, the meridians and the circles of latitude. This is a case of a mapping "by equal parameters (or coordinates)". Incidentally, the theory of mappings of Riemannian spaces originated in cartography which, in turn, is one of its most important fields of application. In the sequel we shall be interested mainly in the case of both manifolds being equipped with a Riemannian metric.

A mapping by equal coordinates between two *Riemannian* spaces can also be given another interpretation: The inverse mapping, from $\tilde{V}_n$ to V_n, can be used to produce a second Riemannian metric in V_n. In the example of geographic maps, we can use on a geographic map, not only its own (plane) Euclidean metric, but also the rules for measuring distances transferred from the sphere that is being mapped. In fact, we shall find it expedient for most purposes to consider, instead of a mapping between two Riemannian spaces V_n and $\tilde{V}_n$, a single manifold with coordinates x^i and equipped with two alternative Riemannian metrics, $g_{ik}(x)$ and $\tilde{g}_{ik}(x)$.

If ξ^i is a tangent vector at (x) then the square of its length is given alternatively by $g_{ik}(x)\xi^i \xi^k$ or by $\tilde{g}_{ik}(x)\xi^i \xi^k$. In particular, let the coordinate system be chosen in such a way that at the point (x) in question, $g_{ik}(x) = \delta_{ik}$ holds. Then the vectors ξ^i that are of unit length relative to the first metric are the vectors of the unit sphere, while the vectors that are of unit length relative to the second metric are those of the ellipsoid $\tilde{g}_{ik}(x)\xi^i \xi^k = 1$. This surface is called the *ellipsoid of Tissot*. It allows a concrete geometric discussion of many important properties of mappings, in particular, also of the distortions of lengths (a "pointwise" property) effected by the mapping (see Fig. 31).

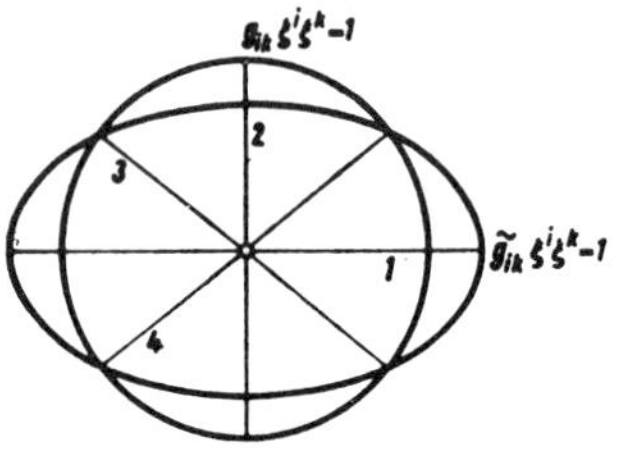

FIG. 31. Tissot's ellipse for $n = 2$. 1, 2 are the directions of principal distortions (directions in which the distortion of lengths assumes extrema); these are mutually orthogonal in both metrics. 3, 4 are the directions of no distortion.

13.2 Mappings with Important Special Properties

We have already encountered various important properties of certain mappings. Let us list some such properties, along with their characterizations in terms of the g_{ik} and $\tilde{g}_{ik}$.

(a) Isometric Mappings. In this case, $\tilde{g}_{ik} = g_{ik}$, and the ellipsoid of Tissot is the unit sphere.

More generally, if the isometric mapping is given by $\tilde{x}^i = \tilde{x}^i(x)$ and not necessarily "by equal coordinates," then

$$g_{ik} = \tilde{g}_{jl} \frac{\partial \tilde{x}^j}{\partial x^i} \frac{\partial \tilde{x}^l}{\partial x^k}.$$

(b) Similarity Mappings or essentially isometric mappings. Here, image and pre-image lengths preserve a constant ratio, and

$$\tilde{g}_{ik}(x) = c \cdot g_{ik}(x).$$

Tissot's ellipsoid, in this case, is a sphere whose radius does not depend on the point (x). If $c \neq 1$, the similarity is "proper."

(c) Isogonal or Conformal Mapping. Here, angles formed by pairs of vectors are preserved under the mapping and, if the mapping is by equal coordinates, then

$$\cos\varphi = \frac{\tilde{g}_{ik}\,\xi^i\,\eta^k}{\sqrt{\tilde{g}_{ik}\,\xi^i\,\xi^k \cdot \tilde{g}_{ik}\,\eta^i\,\eta^k}} = \frac{g_{ik}\,\xi^i\,\eta^k}{\sqrt{g_{ik}\,\xi^i\,\xi^k \cdot g_{ik}\,\eta^i\,\eta^k}}.$$

In particular, if two vectors ξ^i, η^i are orthogonal in the metric $\tilde{g}_{ik}$, then they are also orthogonal in the metric g_{ik}. With $\xi^i = 1$ for $i = 1$ and $\xi^i = 0$ for $i \neq 1$, we therefore find that

$$\tilde{g}_{1k}\eta^k = 0 \qquad \text{and} \qquad g_{1k}\eta^k = 0$$

must hold simultaneously, which they do for $n - 1$ linearly independent η^i. From this, it follows that

$$\tilde{g}_{1k} = c \cdot g_{1k}.$$

The same with any other subscript in place of the subscript 1, and with the same factor c because of the symmetry of the metric tensors g_{ik} and $\tilde{g}_{ik}$. Therefore, a necessary condition for the mapping to be conformal is that

$$\tilde{g}_{ik}(x) = c(x)\, g_{ik}(x).$$

The condition is also sufficient, as is immediately verified. Tissot's ellipsoids for this case are spheres whose radii do depend on the point (x).

(d) Volume-Preserving Mappings. These are characterized by

$$\tilde{g} = g.$$

If $\tilde{g} = c \cdot g$ ($c =$ const.), the mapping is called essentially volume-preserving; all volumes are then multiplied under the mapping by the same constant factor.

For volume-preserving mappings, Tissot's ellipsoid has the same volume as the unit sphere.

The properties discussed so far are "pointwise" properties; they can be expressed solely in terms of the $\tilde{g}_{ik}$ and g_{ik} and can be described in terms of the tangent space, which is done by means of the ellipsoid of Tissot. We pass on next to "infinitesimal" properties, expressed in terms of the derivatives of the fundamental tensors.

(e) Geodesic Mappings. These are mappings that map geodesics onto geodesics. If the mapping is by equal coordinates, it must then satisfy, by Theorem 10.5.1,

$$\tilde{\Gamma}^i_{jk} - \Gamma^i_{jk} = \delta^i_j \varphi_k + \delta^i_k \varphi_j;$$

for the vector φ_k, this implies the condition $\varphi_k = (\tilde{\Gamma}^r_{rk} - \Gamma^r_{rk})/(n+1)$, which, in the present Riemannian case, can be transformed as follows:

$$\Gamma^r_{rk} = \frac{g^{rm}}{2}(g_{rm,k} - g_{rk,m} + g_{km,r}) = \frac{1}{2} g^{rm} g_{rm,k} = \frac{1}{2g}\frac{\partial g}{\partial x^k} = \frac{1}{2}\frac{\partial \log g}{\partial x^k},$$

whence

$$\varphi_k = \frac{1}{2(n+1)} \frac{\partial}{\partial x^k} \log\left(\frac{\tilde{g}}{g}\right).$$

(f) **Affine Mappings.** These are mappings which not only map geodesics onto geodesics but leave natural parametrizations of geodesics invariant. If the mapping is by equal coordinates, this implies that the relations

$$\ddot{x}^i + \tilde{\Gamma}^i_{jk}\,\dot{x}^j\,\dot{x}^k = 0 \qquad \text{and} \qquad \ddot{x}^i + \Gamma^i_{jk}\,\dot{x}^i\,\dot{x}^k = 0$$

are both satisfied with the same parameter t. The symmetry of the Γ then shows the following condition to be necessary and sufficient for the mapping to be affine:

$$\tilde{\Gamma}^i_{jk} = \Gamma^i_{jk}.$$

An affine mapping is evidently also geodesic.

In connection with (e) and (f), note that the conditions obtained for these cases are invariant because the differences $\tilde{\Gamma}^i_{jk} - \Gamma^i_{jk}$ are tensors, even though the Christoffel symbols themselves are not.

Before illustrating all of these definitions by means of examples, let us investigate some interrelations among the various properties. First, it is obvious that an isometric mapping [property (a)] has also all of the remaining properties [that is, (b) through (f)]. A similarity mapping is also conformal and essentially volume-preserving and affine, hence, *a fortiori* geodesic. Hence if we can find a mapping from one space to another that is isometric, or at least a similarity mapping, then we certainly have a mapping that has the other properties also. However, such mappings fail to exist in many cases of practical importance; for example, we know that it is impossible to map any portion of a sphere isometrically into the plane. Even a similarity mapping is not possible in that case, as is shown once again by the *theorema egregium*; for, under a similarity mapping, the Christoffel symbols remain invariant and, therefore, so does the curvature tensor $R^i_{\ jkl}$, while the Gaussian curvature

$$K = g_{2r}\, R^r_{\ 112}/g$$

is multiplied by the positive factor c^{-1}:

$$\tilde{K} = \tilde{g}_{2r}\, \tilde{R}^r_{\ 112}/\tilde{g} = c \cdot g_{2r}\, R^r_{\ 112}/c^2\, g = c^{-1} \cdot K.$$

This proves the asserted impossibility, since $K = 0$ for the plane, whereas $\tilde{K} > 0$ for the sphere. In this case we will therefore have to be satisfied if we can find mappings that have one or the other of the weaker properties (c) — (f).

Exercise 13.2.1. *Prove*: Two Riemannian spaces V_n, $\tilde{V}_n$ of (not necessarily the same) constant curvature can be mapped conformally onto each other. [*Hint.* Use the expression (12.18) for the metric tensor, as it gives the line element in a form that makes the conformal relation to Euclidean space evident.]

Exercise 13.2.2. *Prove* the following formula, of which we made use in (e):

$$\Gamma^r_{rk} = \frac{1}{2}\frac{\partial \log g}{\partial x^k}.$$

[*Hint.* Use $\Gamma^r_{rk} = \frac{1}{2} g^{ri} g_{ri,k}$ and the fact that g^{ri} is the quotient of the principal minor belonging to g_{ri} in $g = \det(g_{ik})$ by g itself. By the product rule of differentiation, we differentiate a determinant by differentiating one row at a time and then adding the results.]

13.3 Some Mappings of the Sphere into the Plane

Before turning to a more general study of the mappings of Riemannian spaces, we shall take a look at some mappings of the sphere that are important for cartography. Let us begin with some mappings that give as faithful as possible an image of the neighborhood of a chosen point, say of the north pole.

(a) Geodesic "Gnomonic" Projection. The sphere is projected from its center onto the tangent plane at the north pole.

This projection maps the geodesics of the sphere — its great circles — onto the straight lines of the plane.

(b) An Equi-Areal Mapping of the Sphere onto Its Tangent Plane at the North Pole. We parametrize the plane by means of polar coordinates r, φ and the sphere by means of the geographic longitude φ and the angle ϑ (see Fig. 32). Setting

$$r = r(\vartheta), \qquad \varphi = \varphi$$

for the mapping to be constructed, we have, for the element of arc length in the plane: $d\tilde{s}^2 = dr^2 + r^2\, d\varphi^2 = r'^2\, d\vartheta^2 + r^2\, d\varphi^2$, and for that of the sphere: $ds^2 = d\vartheta^2 + \sin^2\vartheta\, d\varphi^2$, so that the requirement $\tilde{g} = g$ for an equi-areal mapping yields

$$r^2 r'^2 = \sin^2\vartheta,$$

hence $rr' = \sin \vartheta$ or $r^2/2 = C - \cos \vartheta$. Since $r = 0$ when $\vartheta = 0$, it follows that $C = 1$ for the constant of integration, and the desired equi-areal mapping is given by

$$r = \sqrt{2(1 - \cos \vartheta)} = 2 \sin (\vartheta/2).$$

Figure 33 shows how to construct the mapping. (Recall the theorem on angles subtended by a chord and the circle of Thales.)

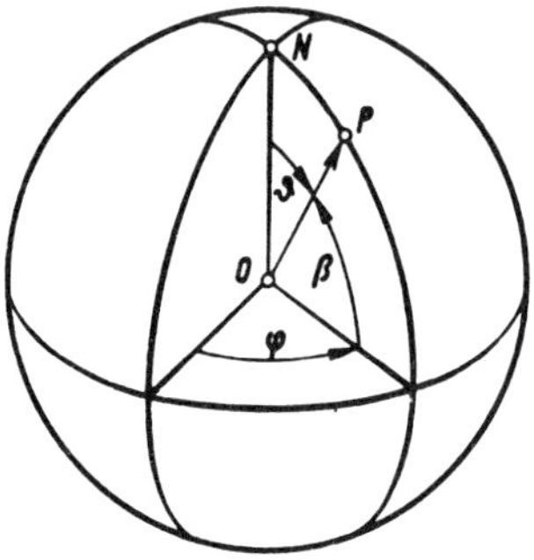

FIG. 32. Geographic coordinates of the point P of the sphere. (φ is the geographic longitude, β is the geographic latitude, $\vartheta = \pi/2 - \beta$ is the angular distance from the pole.)

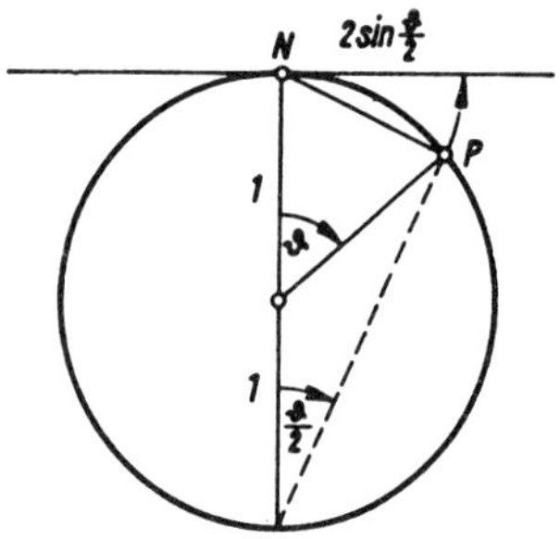

FIG. 33. Radial equi-areal mapping.

(c) The Conformal Stereographic Mapping. Let us set $r = r(\vartheta)$ and $\varphi = \varphi$, as in (b) above, but, this time, for the purpose of finding a conformal mapping. Then the condition $\tilde{g}_{ik} = c \cdot g_{ik}$ leads to the differential equation

$$r' = r/\sin \theta, \qquad \text{hence by integration:} \qquad r = C \cdot \tan (\vartheta/2).$$

For $C = 2$, this yields the stereographic projection of the sphere from its south pole onto the tangent plane at the north pole (see Fig. 34).

Stereographic projection is therefore conformal. For the sake of later applications, we shall show that it also maps circles onto circles: Every circle of the sphere is mapped onto a circle of the plane, straight lines being included among the latter as the images of circles through the south pole. To show this, we construct the circular cone that is tangent to the sphere along a given circle. If we project this cone from the south pole into the plane, its generators are mapped onto a pencil of lines. Since the generators intersect the circle on the sphere orthogonally and since we already know that the mapping is conformal, the image curve of the circle is intersected orthogonally by the lines of the pencil, hence the image curve is a circle.

Conversely, to show that for any circle K of the plane its pre-image on the sphere is a circle of the sphere, we proceed as follows. Let C be the pre-image of K and let K' be a circle on the sphere which at one of its points has contact of the second order with C. Then the image circle of K' has contact of the second order with the circle K and, therefore, coincides with K, whence $K' = C$.

Let us next try to find mappings that give a particularly faithful image of the neighborhood of a great circle, say of the equator. The idea suggests itself of mapping the sphere onto the circular cylinder circumscribed about the sphere along the equator and of then applying this cylinder into the

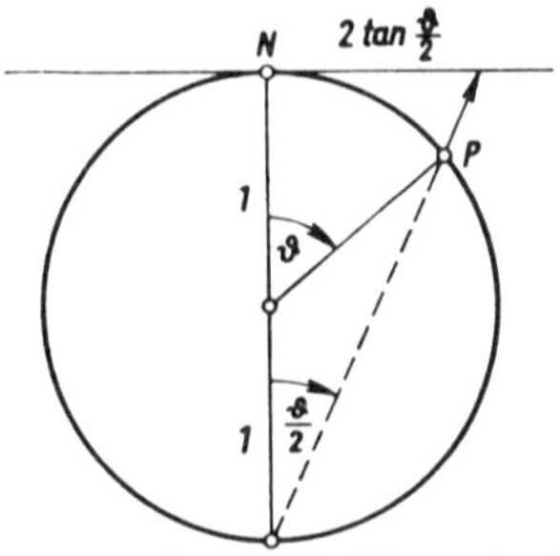

FIG. 34. Stereographic projection.

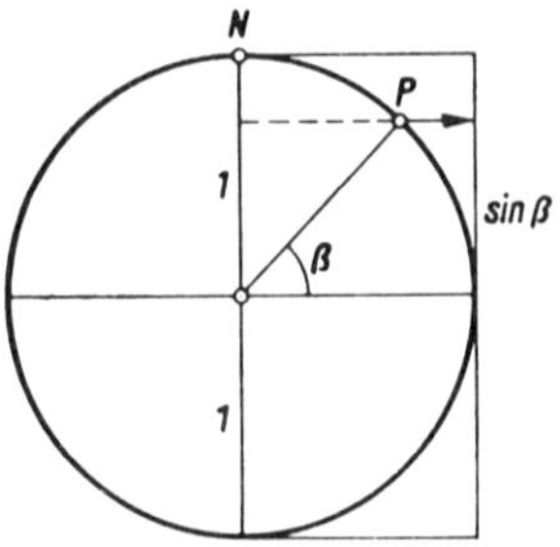

FIG. 35. Equi-areal cylindrical projection.

plane. If geographic coordinates φ, β are used on the unit sphere, its element of arc becomes $ds^2 = d\beta^2 + \cos^2 \beta \, d\varphi^2$, and if the mapping functions are $x = \varphi$, $y = y(\beta)$ then the element of arc of the plane is

$$d\tilde{s}^2 = dx^2 + dy^2 = d\varphi^2 + y'^2 \, d\beta^2.$$

We try first for equi-areal and then for conformal mappings:

(d) Equi-Areal Cylindrical Projection. The condition $\tilde{g} = g$ yields $y' = \cos \beta$, hence $y = \sin \beta$ (the constant of integration is 0 if $y = 0$ corresponds to the equator). Therefore this mapping is produced by orthogonal projection from the north-south axis of the sphere, as we can, of course, also see by elementary geometry (see Fig. 35).

(e) The Isogonal Projection of Mercator. The condition of isogonality leads to the differential equation

$$y' = 1/\cos \beta, \qquad \text{hence} \qquad y = \log \tan(\beta/2 + \pi/4).$$

This conformal map, used by Gerhard Kremer (Mercator) for his map of the world of 1569, has long been the one in use for navigation. A Mercator map using a meridian (circle of longitude) instead of the equator is used in surveying under the name of *Gaussian conformal coordinates*.

13.4 The Incompatibility of Certain Properties of Mappings

In this subsection we show that the most important properties of mappings (being conformal, being volume-preserving, being geodesic) are, in general, mutually exclusive. Specifically, this means that, apart from easily listed exceptional cases, a mapping that is not essentially isometric can have at most one of the three properties listed above.

Theorem 13.4.1. *If a mapping is both conformal and (essentially) volume-preserving, then it is also (essentially) isometric.*

Proof. Isogonality implies $\tilde{g}_{ik} = \lambda(x)g_{ik}$, (essential) preservation of volume implies $\tilde{g} = c \cdot g$. The former condition yields $\tilde{g} = \lambda^n(x)g$, hence the latter now gives $\lambda(x) = \sqrt[n]{c} = \text{const.}$, and this constant is 1 if volumes are literally (not only essentially) preserved.

The next result is less evident:

Theorem 13.4.2. *If a mapping is both conformal and geodesic, then it is a similarity mapping, hence essentially isometric.*

Proof. Since the mapping is geodesic, the difference of the Christoffel symbols satisfies the equation given under (e) in §13.2. We shall transform this difference by also using the isogonality condition $\tilde{g}_{ik} = c(x)g_{ik}$. First, $\tilde{g}^{ik} = (1/c)g^{ik}$ and $\tilde{g}_{ik,j} = cg_{ik,j} + c_{,j}\, g_{ik}$. It follows that, for conformal mappings,

$$\tilde{\Gamma}^i_{jk} - \Gamma^i_{jk} = \frac{1}{2c} g^{ir}(g_{rj}\, c_{,k} - g_{jk}\, c_{,r} + g_{rk}\, c_{,j}) = \delta^i_j\, d_{,k} + \delta^i_k\, d_{,j} - g^{ir}\, g_{jk}\, d_{,r} \qquad (*)$$

where $d = \log\sqrt{c}$. Now, for fixed i, choose $j = k \neq i$; then, using also (e) of §13.2, we obtain $g^{ir}\, d_{,r} = 0$ and, from this, by inner multiplication with g_{ki}, we have $d_{,k} = 0$, from which the assertion follows.

If a mapping is both geodesic and volume-preserving, equally strong conclusions cannot be drawn, without further assumptions:

Theorem 13.4.3. *If a mapping is both volume-preserving and geodesic, then it is affine.*

Proof. If we substitute $\tilde{g} = c \cdot g$ (the preservation-of-volumes condition) into (e) of §13.2, we obtain the equality of the Christoffel symbols, hence the assertion, by (f) of §13.2.

In the case of dimensionality 2, the conclusion implies stronger results than are spelled out in Theorem 13.4.3:

Theorem 13.4.4. *An affine mapping from one two-dimensional space onto another is either a similarity mapping, or else the two spaces are Euclidean, in which case the mapping is an affinity in the ordinary, elementary sense.*

Proof. To prove this, it is easiest to use geodesic parallel displacement. Under an affine mapping, by equal coordinates, the Christoffel symbols are the same, hence geodesic parallelism is preserved in passing from one metric to the other. At any fixed point, consider the ellipses $g_{ik}\,\xi^i\,\xi^k = 1$ and $\tilde{g}_{ik}\,\xi^i\,\xi^k = 1$. Under the holonomy group of the common parallel displacement, each of the two ellipses is mapped onto itself. Hence they must be similar and, in fact, homothetic, unless the group consists only of its identity element. In the latter case, both metrics are Euclidean, while in the former we must have $\tilde{g}_{ik} = c \cdot g_{ik}$, where the factor c must be independent of the point because of the isometric character of parallel displacement.

For the case of the theory of surfaces, we can summarize the preceding theorems as follows, making use of the fact that developables are the only surfaces with a Euclidean metric:

Theorem 13.4.5. *Let there be given a mapping that is not essentially isometric, between two surfaces of which at least one is not a developable surface. Then the mapping can have at most one of the three properties of being conformal, area-preserving, or geodesic.*

Another look at the proof of Theorem 13.4.4 shows that it will go through, even for $n > 2$, provided only that the holonomy group is "sufficiently big," in the sense, say, that every direction at the point can be mapped onto any other direction at the point by an element of the group. Spaces of this sort are called *irreducible.* Hence we have

Theorem 13.4.6. *If two Riemannian spaces of which one is irreducible are connected by an affine mapping, then this mapping is a similarity mapping. If the spaces are connected by a mapping that is not a similarity mapping, then it can have at most one of the three properties of being conformal, volume-preserving, and geodesic.*

The proof is as above.

Exercise 13.4.7. *Prove* Theorem 13.4.4 analytically, without using the holonomy group. [*Hint.* Use the orthogonal net common to the two metrics. The proof is carried out in *Jahresber. DMV* **60**, 93–96 (1957).]

Exercise 13.4.8. If $n > 2$, then it does not follow from $\tilde{\Gamma}^i_{jk} = \Gamma^i_{jk}$ that the two metrics are Euclidean or that the mapping is a similarity. To be sure, neither of the two metrics is irreducible in this case, by Theorem 13.4.6. Give examples! (*Hint.* Set $g_{ik} = g_i\,\delta_{ik}$, $\tilde{g}_{ik} = g_{ik}$ for $i, k < n$ and $\tilde{g}_{nk} = c \cdot g_n\,\delta_{nk}$ with $c \neq 1$.)

13.5 Geodesic Mappings Into Euclidean Space

Geodesic mappings are relatively rare. It is not possible to map any V_n geodesically into any other $\tilde{V}_n$. On the other hand, volume-preserving mappings are always easy to find and, for $n = 2$, conformal mappings can always be found in many ways.

Theorem 13.5.1. (Beltrami, 1865.) *A V_2 can be mapped geodesically into the plane if, and only if, V_2 is of constant curvature.*

Proof. We choose common coordinates for the two spaces in such a way that $\tilde{\Gamma}^i_{jk} = 0$ holds in the plane. Its geodesic pre-image V_2 then has Christoffel symbols Γ that satisfy, by (e) of §13.2,

$$\Gamma^2_{11} = \Gamma^1_{22} = 0, \qquad \Gamma^1_{11} = 2\,\Gamma^2_{12}, \qquad \Gamma^2_{22} = 2\,\Gamma^1_{21}. \tag{13.2}$$

Hence the Gaussian curvature K of V_2 satisfies

$$K\,g_{11} = \Gamma^2_{12}\,\Gamma^2_{12} - \Gamma^2_{12,1} \qquad \text{and} \qquad K\,g_{12} = \Gamma^1_{21}\,\Gamma^2_{12} - \Gamma^2_{12,2}, \tag{13.3}$$

as well as the equations obtained from these by interchanging 1 with 2. Now we differentiate equations (13.3) with respect to x^2 and with respect to x^1, subtract, and substitute from (13.3), obtaining

$$g_{11}\,K_{,2} - g_{12}\,K_{,1} + K(g_{11,2} - g_{12,1}) = K(g_{11}\,\Gamma^1_{12} - g_{12}\,\Gamma^2_{12})$$

and from this, after substituting the explicit expressions for the Christoffel symbols,[1]

$$g_{11}\,K_{,2} - g_{12}\,K_{,1} = 0.$$

Similarly one obtains the equation resulting from an interchange of the indices 1 and 2. Hence $g^{ij}\,K_{,j} = 0$ and therefore $K = \text{const.}$, which proves the necessity of the condition.

To prove its sufficiency, let V_2 be a space of constant curvature K; we shall explicitly construct a geodesic mapping of V_2 into the Euclidean plane. If $K = 0$, nothing remains to be proved. If $K > 0$, the space V_2 is

[1] It is easiest to substitute, for the derivatives of the g_{ik} on the left-hand side, from Ricci's lemma, and then to make use of (13.2).

locally isometric to a sphere (by §12.3) and the gnomonic projection of (a), §13.3, can therefore be used as the desired geodesic mapping. If $K < 0$, we make use of the Poincaré half-plane (§12.6), onto which all V_2 of constant negative curvature can be mapped by (local) similarities (see §12.3) and hence geodesically.

Besides the Poincaré half-plane E, with boundary line l, we introduce further auxiliary surfaces in Euclidean 3-space: A unit sphere tangent to E, with the point N on this sphere being diametrically opposite to the point of tangency; and a plane E' parallel to l and perpendicular to E (see Fig. 36). Now we project E stereographically onto (part of) the sphere, from N,

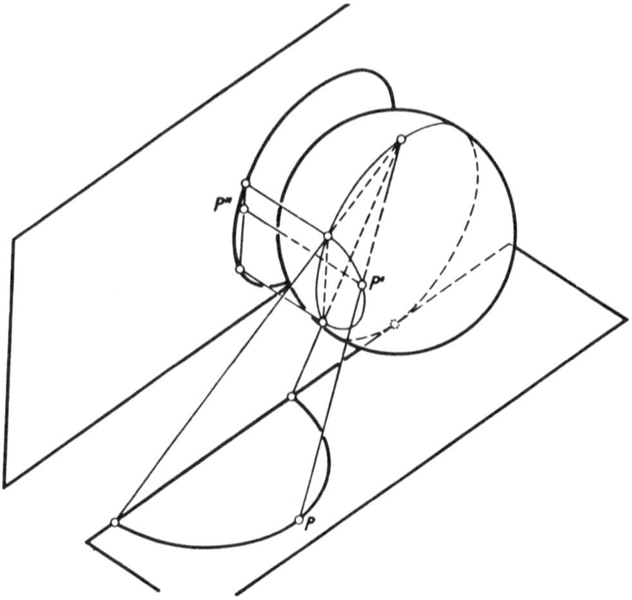

FIG. 36. Geodesic mapping of the Poincaré half-plane.

carrying over to the sphere the non-Euclidean metric of E. By what was proved in (c) of §13.3, the geodesics of the Poincaré half-plane — that is, the circles perpendicular to l — are mapped into circles of the sphere that are perpendicular to the meridian through N which is tangent to l. Next, we use parallel projection perpendicular to E', from the sphere into E'. This maps the circles just described onto all straight line segments within a unit circle of E'. We have thus constructed a geodesic mapping of the entire hyperbolic plane onto the interior of a unit circle of the Euclidean plane.

We can restate Theorem 13.5.1 in the following more general formulation:

Theorem 13.5.2. *A Riemannian space V_2 can be mapped geodesically onto a space $\tilde{V}_2$ of constant curvature if and only if V_2 itself is of constant curvature.*

Corresponding theorems, incidentally, also hold for any dimension $n > 2$.

The image of the hyperbolic plane in the interior of a unit circle is of great importance, as the so-called *projective model of the hyperbolic plane.* These non-Euclidean metrics, as the only ones whose geodesics are the straight lines, occur in projective geometry, as is well known, as metrizations of the projective plane.

We obtain another important model of the hyperbolic plane by projecting the sphere into the plane E' not orthogonally but stereographically, from the point Q of the sphere whose tangent plane is parallel to E' (and on the side of E'). The image is once again the interior of a circle C of E' and the geodesics are the circular arcs inside C that are perpendicular to C. This model is conformal, as follows from (c) of §13.3.

Exercise 13.5.3. U. Dini [*Ann. mat.* (2) **3** (1869)] has determined all pairs of surfaces that can be mapped onto each other geodesically. *Prove*: In suitable parameters, the condition for this is that

$$g_{ik} = (U_1 - U_2)\delta_{ik}, \qquad \tilde{g}_{ik} = \left(\frac{1}{U_2} - \frac{1}{U_1}\right)\frac{\delta_{ik}}{U_i},$$

where U_i depends only on u^i.

(*Hint.* Choose a common system of parameters orthogonal on both surfaces, so that $g_{12} = \tilde{g}_{12} = 0$, and show that under a geodesic mapping, the ratio

$$\frac{\tilde{g}_{11}}{\tilde{g}_{22}^2} : \frac{g_{11}}{g_{22}^2}$$

is independent of u^1; correspondingly also, with 1, 2 interchanged.)

Exercise 13.5.4. *Prove*: The surfaces for which $g_{ik} = (U_1 - U_2)\delta_{ik}$ include all surfaces of revolution and all general helicoids.

[*Hint.* Using geographic parameters for a surface of revolution, and the parameter transformation

$$\bar{u}_1 = \int \frac{du^1}{\sqrt{G(u^1)}},$$

we have

$$ds^2 = (du^1)^2 + G(u^1)(du^2)^2 = G(u^1)((d\bar{u}^1)^2 + (d\bar{u}^2)^2)$$

and we know from Bour's theorem (6.5.1) that the general helicoids are isometric to surfaces of revolution.]

13.6 Similarity Mappings and Affine Mappings of a Space onto Itself

We have characterized the spaces of constant curvature as those that can be mapped geodesically into Euclidean space. We shall follow this up with an interesting characterization of Euclidean spaces, namely: They are the only ones that allow similarity mappings other than isometries onto themselves. We begin with a very simple lemma about similarity mappings between two spaces.

Lemma. *Let φ be the angle between a vector ξ at P and a vector η at P that results from ξ by parallel displacement along a path beginning and ending at P. Let $\tilde{\varphi}$ be the corresponding angle in a space $\tilde{V}_n$ mapped onto V_n by a similarity mapping. Then $\tilde{\varphi} = \varphi$.*

The proof follows immediately from the equality of the Christoffel symbols and the resulting correspondence of the parallel displacements.

Theorem 13.6.1. *If a complete Riemannian space allows a proper similarity mapping onto itself, then it is a Euclidean space.*

Proof. We show first that such a similarity mapping $P \to P'$ must have a fixed point, that is, a point mapped onto itself. For, to begin with, all lengths are multiplied under the mapping by a factor $c \neq 1$, and we may assume $0 < c < 1$ (for otherwise, the inverse mapping may be considered instead). Thus the mapping reduces all lengths in the fixed ratio $1 : c$, that is, it is a contraction mapping. By a principle of S. Banach used frequently in analysis, such a mapping has exactly one fixed point, as can be seen as follows: Consider a point P, its image point P', the image point P'' of P', and so forth. For any points P, Q, let $L(P, Q)$ be the infimum of the lengths of all arcs that join P and Q. Then, clearly,

$$L(P', P'') = c \cdot L(P, P'), \qquad L(P'', P''') = c \cdot L(P', P'') = c^2 \cdot L(P, P'), \text{ etc.}$$

Hence the sequence of image points of P is a Cauchy sequence, since $c < 1$, and, on account of the completeness of the space, it converges to a point O which must be a fixed point (because its image O' is the limit of the sequence $P', P'', P''', \ldots$ as O itself is the limit of the successive image points $P, P', P'', \ldots$). There cannot be a second fixed point F, with $F' = F$, since we would otherwise have a contradiction by $cL(O, F) = L(O', F') = L(O, F)$.

We prove next that the holonomy group at O consists only of the identity element, which is equivalent to the statement that the space is Euclidean.

Let C be a curve that begins and ends at O, let ξ be a vector at O, and let φ be the angle between ξ and the vector η that is obtained from ξ by parallel displacement around C. We map all of this by means of our contraction mapping, obtaining an angle φ' between ξ' and the vector η' obtained from ξ' by parallel displacement around C'. By the lemma, we have $\varphi' = \varphi$. Repetitions of this process lead to image curves $C^{(n)}$ lying in an arbitrarily small neighborhood of O, so that the corresponding angle $\varphi^{(n)}$ is also as small as we wish. On the other hand, $\varphi^{(n)} = \ldots = \varphi' = \varphi$, hence $\varphi = 0$, and the proof is complete.

Affine mappings of a Riemannian space onto itself are also rare. We confine ourselves to irreducible spaces, which include, in particular, the spaces of constant curvature $K \neq 0$ (*Exercise*: Prove this statement!) and we have, in this case, the following consequence of Theorem 13.4.6:

Theorem 13.6.2. *Any affine mapping of an irreducible Riemannian space onto itself must be an isometry.*

Thus, in particular, the only affine mappings onto itself that a space of constant curvature $K \neq 0$ allows are isometric mappings — unlike the Euclidean case ($K = 0$) in which, as we know, there exist affine mappings that are not isometries.

This characterizes Euclidean geometry among the classical "non-Euclidean" geometries.

Exercise 13.6.3. The hypothesis of completeness in Theorem 13.6.1 is essential, as is shown by the following spiral surfaces, investigated in this connection by H. Lenz. The mappings

$$S(t) \underset{\text{def}}{=} e^t \begin{pmatrix} \cos t & \sin t & 0 \\ -\sin t & \cos t & 0 \\ 0 & 0 & 1 \end{pmatrix}$$

constitute a group of similarities of three-dimensional Euclidean space onto itself that leave the x_3-axis invariant. Let $\mathfrak{x}(u_1) = (x(u_1), 0, z(u_1))$ be a nonrectilinear curve in the (x_1, x_3)-plane. *Prove* that

$$\mathfrak{x}(u_1, u_2) \underset{\text{def}}{=} S(u_2)\mathfrak{x}(u_1)$$

is a surface whose Gaussian curvature K does not vanish identically and which allows similarity mappings into itself. (*Hint.* Show that $g_{ik}(u_1, u_2 + c) = e^{2c} \cdot g_{ik}(u_1, u_2)$, from which it follows that the mappings $u_2 \to u_2 + c$, $u_1 \to u_1$ are similarity mappings of the surface. Apart from a positive factor, K equals $xz'(z''x' - x''z')$, and this does not vanish identically, except for straight lines.)

Exercise 13.6.4. Give a precise argument for the last step of the proof of Theorem 13.6.1. [*Prove*: Given $\varepsilon > 0$, there exists $\delta(\varepsilon) > 0$ such that, for any path of length $l \leq \delta(\varepsilon)$ beginning and ending at P, parallel displacement around that path of a vector ξ^i yields a change in angle of no more than ε. *Hint*. The function $\xi^i(s)$ is a solution of

$$\frac{d\xi^i}{ds} = -\Gamma^i_{jk}\,\xi^j\,\frac{dx^k}{ds}$$

and, by the mean value theorem, we have

$$|\xi^i(l) - \xi^i(0)| \leq l \cdot \left|\frac{d\xi^i}{ds}\right|,$$

with the right-hand side to be evaluated at an intermediate value of s. Use the differential equation, also the fact that ξ^j and dx^k/ds are unit vectors, also the continuity and therefore boundedness of the Γ in a neighborhood of P.]

13.7 Conformal Mapping of Two-Dimensional Spaces

We have already constructed explicitly a number of conformal mappings between Riemannian spaces: In §12.5 we constructed conformal mappings from spaces of constant curvature into Euclidean space (and, therefore, also into each other); again, Exercise 13.5.4 shows that rotation surfaces and general helicoids can be conformally mapped into the plane. And, in fact, it is always an easy matter to obtain explicitly any conformal mapping that is needed in differential geometry. The theorem that *any* two-dimensional Riemannian space can be mapped (locally) conformally into the plane is therefore of greater interest for analysis than for geometry. (The theorem applies to Riemannian spaces of class C^3 and can even be strengthened somewhat. See A. Korn, "Zwei Anwendungen der Methode der sukzessiven Annäherungen," pp. 215–229, Schwarz-Festschrift, 1914. L. Lichtenstein, in 1916, strengthened the theorem to twice-differentiable surfaces satisfying a Hölder condition.) Although this existence proof does not fit into our systematic development of differential geometry, we shall still reproduce here the historically important and interesting theorem of Gauss (1822) to the effect that any analytic surface can be mapped conformally into the plane. (C. F. Gauss, Allgemeine Auflösung der Aufgabe, die Teile einer gegebenen Fläche ... so abzubilden, daß die Abbildung dem Abgebildeten in den kleinsten Teilen ähnlich wird. "Works," Vol. IV, p. 189; 2nd printing, Göttingen, 1880.)

Theorem 13.7.1. (Gauss, 1822.) *Every analytic surface can be mapped conformally into the plane.*

Proof. Clearly, the problem is simply that of introducing "isothermal" parameters on the surface, that is, parameters in terms of which $\bar{g}_{ik} = E(\bar{u}^j)\delta_{ik}$. We start with any parametrization, in terms of which we have $ds^2 = g_{ik}\, du^i\, du^k$, where the g_{ik} are, by hypothesis, analytic functions, i.e., have power series expansions. It follows that — and this turns out to be the essential point to Gauss' analyticity assumption — these functions can be extended to complex arguments, within the domain of convergence of their power series. We decompose $g_{ik}\, du^i\, du^k$ into two complex conjugate linear factors and use Theorem B of §3.3 to construct a complex coordinate transformation $\bar{u}^i(u^k)$ such that the parametric curves $\bar{u}^i =$ const. satisfy the condition $d\bar{s} = 0$, which implies $\bar{g}_{11} = \bar{g}_{22} = 0$. (Curves of this sort can exist only in the complex domain; they are called *ametric*, or — an older name — *isotropic*.) Thus $ds^2 = 2F\, d\bar{u}^1\, d\bar{u}^2$. From Theorem B of §3.3, it follows that the parameters $\bar{u}^i$ are complex conjugates. Hence there are real functions v^i such that

$$d\bar{u}^1 = dv^1 + i\, dv^2$$

$$d\bar{u}^2 = dv^1 - i\, dv^2.$$

Introducing this into ds^2, we obtain

$$ds^2 = 2F[(dv^1)^2 + (dv^2)^2],$$

which shows that for v^i real, F itself is real and positive. The element of arc is thus exhibited in isothermal form. The factor $2F$ is an analytic function of v^i.

Conformal mappings and their continuations — thus, "in the large" — are studied in detail in the theory of functions.

Exercise 13.7.2. Let x, y be Cartesian coordinates in the plane. *Prove*: The mapping $(x, y) \to (u(x, y), v(x, y))$ from one region of the plane to another is conformal if and only if the differential equations of Cauchy and Riemann apply:

$$u_x = \pm v_y \quad \text{and} \quad u_y = \mp v_x.$$

[*Hint*. Isogonality is equivalent to

$$du^2 + dv^2 = g(x, y)(dx^2 + dy^2),$$

and this is, in turn, equivalent to

$$\text{(a)}\quad u_x{}^2 + v_x{}^2 = g = u_y{}^2 + v_y{}^2 \quad \text{and} \quad \text{(b)}\quad u_x u_y + v_x v_y = 0;$$

now (b) is equivalent to $u_x = hv_y$ and $hu_y = -v_x$, and (a) then requires that $h = \pm 1$.]

Note further that the Cauchy-Riemann differential equations with the choice of the upper signs lead to analytic functions $w = w(z)$ of the complex variable z if we introduce $z = x + iy$ and $w = u + iv$. The lower signs lead to functions $\bar{w} = u - iv$. Hence, all isogonal mappings of the plane are represented by complex analytic functions of a complex variable — (strictly) conformal mappings — and by the conjugates of such functions — "anticonformal" mappings.

Exercise 13.7.3. *Prove* the following result of Liouville (1850): A conformal mapping of space maps spheres onto spheres. (By contrast, a conformal mapping of the plane does not, in general, map circles onto circles.) (*Hint.* On account of isogonality, any triply orthogonal system of surfaces — *cf.* § 6.8 — is mapped onto another such system. Take a sphere to be one member of such a system, and consider any given curve on that sphere to be the curve of intersection with another surface of the system, which Theorem 6.8.2 entitles us to do. By Theorem 6.8.2, the image of that curve is itself a line of curvature on the image of the sphere. Thus all curves on the image surface are lines of curvature, hence the image surface is a sphere.)

Exercise 13.7.4. *Prove* that similarity transformations in the elementary sense — i.e., rigid motions followed by multiplication of lengths by a fixed factor — are similarity mappings, as are also inversions. (With 0 being the Cartesian coordinate origin, an *inversion* $\mathfrak{x} \to \mathfrak{y}$ relative to 0, or with center 0, is a mapping of the form

$$\mathfrak{y} = c^2 \cdot \frac{\mathfrak{x}}{\mathfrak{x}^2};$$

this is also called a *reflection* in the sphere of radius c and center at 0.) [*Hint* for the case of inversions: Form $d\mathfrak{y} = c^2\, d\mathfrak{x}/\mathfrak{x}^2 - \mathfrak{x}c^2\, 2(\mathfrak{x}\, d\mathfrak{x})/(\mathfrak{x}^2)^2$ and show that $d\mathfrak{y}^2 = \lambda\, d\mathfrak{x}^2$.]

Exercise 13.7.5. *Prove* that every isogonal mapping of three-dimensional space can be compounded from the types listed in Exercise 13.7.4. (*Hint.* Consider a triply orthogonal system of spheres; at a point $\mathfrak{x}_0$ the three spheres through that point are tangent to three mutually perpendicular planes. An isogonal mapping $A: \mathfrak{x} \to \mathfrak{y}$ maps the three spheres onto a similar triple through the image point $\mathfrak{y}_0$ of $\mathfrak{x}_0$, by Exercise 13.7.3. Let I and J be two inversions, with centers at $\mathfrak{x}_0$ and at $\mathfrak{y}_0$, respectively. These inversions map the triply orthogonal spheres onto triply orthogonal planes. The mapping $J^{-1}AI$ maps the one system of planes onto the other, and it is a composite of isogonal mappings, therefore itself isogonal; in fact, it is a similarity transformation.)

13.8 Minimal Surfaces

Let us determine those surfaces of Euclidean 3-space for which Gauss' spherical mapping by parallel normals (§ 5.3) is a conformal mapping. The parametrizations of the surface $\mathfrak{x}(u^i)$ and of its spherical image $\mathfrak{N}(u^i)$ are the same, therefore the conformality condition of (c), § 13.2, becomes

$$\mathfrak{N}_i \mathfrak{N}_k = c(u^i)\, \mathfrak{x}_i\, \mathfrak{x}_k. \tag{13.4}$$

Using a parametrization in terms of the lines of curvature, we know from Rodrigues' formulas (Exercise 5.3.2) that $\mathfrak{N}_i = - \kappa_{(i)} \mathfrak{x}_i$, hence (13.4) gives $\kappa_{(1)}^2 = \kappa_{(2)}^2$ and therefore either $\kappa_{(1)} = \kappa_{(2)}$ or $2H = \kappa_{(1)} + \kappa_{(2)} = 0$. In the former case, we have a sphere; in the latter, a minimal surface (see §6.6). The argument is reversible, showing that in these cases the spherical mapping of Gauss is indeed conformal.

In this context, let us return once more to the study of minimal surfaces. We shall only consider parts of surfaces that have no flat points, which excludes, in particular, the trivial case of the plane. Then we always have $K < 0$, since $\kappa_{(1)} = - \kappa_{(2)}$ implies $K \leqq 0$ and since $K = 0$ is possible only if both principal curvatures vanish, that is, at a flat point. (The reader will convince himself easily that, if $K \neq 0$, then the spherical mapping is locally one to one. This follows from Gauss' interpretation of K, see §5.3.)

Minimal surfaces have been the subject of extensive and thorough studies. Their theory makes use of very special properties of just this class of surfaces and thus does not contribute much to the general theory of surfaces; but it contains many beautiful results and has very close connections with complex function theory. These connections are due, for one thing, to the conformal character of the Gaussian spherical mapping and, for another, to the analyticity of all minimal surfaces. This latter property we shall prove now.

Let u^i be isothermal parameters on the minimal surface. (We do not have to call on the general existence theorem for such parametrizations, not proved in this book. Instead, we can use the conformal spherical mapping to transfer an isothermal net on the sphere onto an isothermal net on the minimal surface.)

Since $H = 0$, we have — using Beltrami's operator of §11.7, Exercise 11.7.1. —

$$\Delta \mathfrak{x} = 0,$$

which becomes, in isothermal parameters,

$$\mathfrak{x}_{11} + \mathfrak{x}_{22} = 0.$$

Hence each of the three components of the position vector is a solution of Laplace's equation and, therefore, by a theorem of potential theory, each is the real part of an analytic function of the complex variable $u = u^1 + i \cdot u^2$, so that

$$\mathfrak{x} = \operatorname{Re} \mathfrak{z},$$

where

$$\mathfrak{z}(u) = \mathfrak{x}(u^1, u^2) + i\,\mathfrak{y}(u^1, u^2).$$

Thus we see that in isothermal parameters, minimal surfaces of class C^2 are even analytic.

Conversely, let $\mathfrak{z} = \mathfrak{z}(u)$ be a complex analytic vector function. For its real part $\mathfrak{x}$ to represent a minimal surface referred to isothermal coordinates, the conditions $\mathfrak{x}_1{}^2 = \mathfrak{x}_2{}^2$, $\mathfrak{x}_1\,\mathfrak{x}_2 = 0$ must be satisfied. Since $\mathfrak{z}' = \mathfrak{x}_1 - i\mathfrak{x}_2$ by the Cauchy-Riemann differential equations, these conditions are equivalent to

$$\mathfrak{z}'^2 = 0.$$

Thus we have the following result:

Theorem 13.8.1. *The problem of determining all minimal surfaces* $\mathfrak{x}(u^1, u^2)$ *in isothermal coordinates* u^i *is equivalent to the problem of determining all complex analytic functions* $\mathfrak{z} = \mathfrak{z}(u)$ *of* $u = u^1 + i \cdot u^2$ *for which* $\mathfrak{z}'^2 = 0$ *and setting*

$$\mathfrak{x}(u^1, u^2) = \operatorname{Re} \mathfrak{z}(u).$$

Such functions $\mathfrak{z}(u)$ have sometimes been interpreted as curves in a complex three-dimensional space referred to the complex parameter u. The condition $\mathfrak{z}'^2 = 0$ is then expressed by saying that the length of this curve is 0 (because of the analogy between $\mathfrak{z}'^2$ and the square of arc length in the real case), or that the curve is *ametric* or *isotropic*. We shall not follow up this analogy any further, since it is not of much interest for geometry, being useful only for introducing a pseudo-geometric terminology to describe purely analytic matters.

Following Weierstrass, we give the following explicit representation of minimal surfaces, which is even more systematic. From $\mathfrak{z}'^2 = 0$, it follows that there exist two analytic functions $g(u)$, $h(u)$ such that the components of $\mathfrak{z}'$ satisfy

$$z^{1\prime} = g^2 - h^2, \qquad z^{2\prime} = i(g^2 + h^2), \qquad z^{3\prime} = 2gh, \tag{13.5}$$

whence the components of the real part are

$$x^1 = \operatorname{Re} \int g^2\left(1 - \frac{h^2}{g^2}\right) du$$

$$x^2 = \mathrm{Re} \int i\, g^2 \left(1 + \frac{h^2}{g^2}\right) du$$

$$x^3 = \mathrm{Re} \int 2 \frac{h}{g} g^2\, du.$$

We assume for the moment that $gh' - g'h \neq 0$ and we set $h/g = \xi$, $g^4/(gh' - g'h) = F(\xi)$; then

$$x^1 = \mathrm{Re} \int (1 - \xi^2) F(\xi) \, d\xi, \qquad x^2 = \mathrm{Re} \int i(1 + \xi^2) F(\xi)\, d\xi,$$

$$x^3 = \mathrm{Re} \int 2\xi F(\xi)\, d\xi$$

and, setting $F = f'''$, we can write, without using integral signs,

$$\begin{aligned} x^1 &= \mathrm{Re}\,(f''(1 - \xi^2) + 2\xi f' - 2f) \\ x^2 &= \mathrm{Re}\,(i(1 + \xi^2)f'' - 2i\xi f' + 2if) \\ x^3 &= \mathrm{Re}\,(2\xi f'' - 2f'). \end{aligned} \tag{13.6}$$

In this way, every analytic function whose third derivative does not vanish identically yields a minimal surface in isothermal parameters. What is more, this construction yields all minimal surfaces with the exception of planes, since the conditions $g \neq 0$ and $gh' - g'h \neq 0$ exclude only planes [$g = 0$ gives $z^3 = \text{const.}$, hence $x^3 = \text{const.}$, and $gh' - g'h = 0$ gives $h/g = 0$ which, by (13.5), again leads only to planes].

The connections with complex function theory allow for many applications. For example, all minimal surfaces being analytic, we can introduce ametric parameters $\bar{u}^i$ in such a way (see the proof of Theorem 13.7.1) that (a) $\bar{g}_{11} = \bar{g}_{22} = 0$ and, furthermore, by $0 = \bar{H} = \bar{L}_{12}/\bar{g}_{12}$, also

$$\bar{L}_{12} = \mathfrak{N}\mathfrak{x}_{12} = 0,$$

and, since differentiation of (a) yields $\mathfrak{x}_1 \mathfrak{x}_{12} = \mathfrak{x}_2 \mathfrak{x}_{12} = 0$, it follows that $\mathfrak{x}_{12} = 0$. Use these facts to solve the following problem:

Exercise 13.8.2. *Prove*: An analytic surface is a minimal surface if, and only if, it can be represented in the form

$$\mathfrak{x}(\bar{u}^1, \bar{u}^2) = \mathfrak{a}(\bar{u}^1) + \mathfrak{b}(\bar{u}^2),$$

where $\mathfrak{a}$ and $\mathfrak{b}$ are ametric curves in their (complex) parameters. [Surfaces of the form $\mathfrak{a}(u^1) + \mathfrak{b}(u^2)$ are also called *translation surfaces*, since they can be obtained by translating the curve $\mathfrak{b}$ along the curve $\mathfrak{a}$. Following Sophus Lie (1879), we can therefore say that minimal surfaces are translation surfaces with ametric generators.]

§ 14. Riemannian Spaces and Analytical Dynamics

14.1 Phase Space and the Kinematic Element of Arc

Riemannian geometry and the tensor calculus have many applications in mechanics and physics. A textbook of differential geometry is not the place where these applications can be followed up in detail. Here we shall confine ourselves to a brief treatment of a small branch of these applications, namely, the mechanics of systems with finitely many degrees of freedom. This example is particularly closely connected with Riemannian geometry and can also lead to a better insight into the purely geometric facts, just as the kinematic interpretations in the elementary theory of curves and surfaces are a help toward a more concrete comprehension of concepts and results. Besides, the parts of analytical mechanics that we shall present here hardly require any new knowledge beyond the differential geometry that has already been traversed, and this is an advantage that we could not count on if we were to consider applications to other branches of physics.

The possible states of a dynamical system can be represented by the points of a differentiable manifold which is called the *phase space* (or *configuration space*). A coordinate system in such a space consists of the "generalized coordinates" q^i. The dimension n of the space is the number of the *degrees of freedom*. We give a few simple examples.

$n = 1$: A mass point allowed to move on a given curve. Here, the arc length s of the curve is a suitable choice for the single coordinate q^1. A particular example is a pendulum swinging in a fixed vertical plane.

$n = 2$: A mass point allowed to move on a given surface. As coordinates q^1, q^2, any parameters for the surface are suitable. A particular example is a pendulum, not constrained to a vertical plane. Another example is a double pendulum swinging in a fixed vertical plane; in this example, suitable coordinates q^i are, for example, the two angles made with the vertical.

$n = 3$: A mass point in space. Also, a rigid body free to rotate about one of its points (which is held fixed), for instance, a spinning top.

Another interesting example is that of two mass points in the plane rigidly connected by a rod of unit length. If (x_1, x_2) and (y_1, y_2) are the Cartesian coordinates of the two points, we therefore have the restraining equation ("*restraint*")

$$(x_1 - y_1)^2 + (x_2 - y_2)^2 = 1. \tag{14.1}$$

Hence the possible states (positions) of the system are given by all those points of four-dimensional space (with coordinates x_1, x_2, y_1, y_2) that lie on the three-dimensional hypersurface defined by (14.1). Any parametrization q^1, q^2, q^3 of this hypersurface yields suitable coordinates for the phase space; for instance, we can set $q^1 = x_1$, $q^2 = x_2$, $q^3 = y_1$, and then express y_2 in terms of these coordinates by means of (14.1). Alternatively, we can choose q^1 and q^2 as before but then use for $\bar{q}^3$ the angle ϑ between the rod and the horizontal, so that

$$y_1 = x_1 + \cos\vartheta, \qquad y_2 = x_2 + \sin\vartheta.$$

$n = 4$: Two mass points moving on a given surface; for instance (the centers of gravity of) sun and earth moving in the plane of the ecliptic.

$n = 5$: Two rigidly connected mass points (a "rod") in (three-dimensional) space.

$n = 6$: Two free mass points in space. Also, a rigid body free to move in space.

The list can be continued at will. For every positive integer n there is a dynamical system with n degrees of freedom. Suppose that the system consists of N points in 3-space; these points have altogether $3N$ coordinates, and these coordinates will, in general, be subject to restraints of the form

$$F_i(x_1, x_2, x_3, y_1, \ldots) = 0 \tag{14.2}$$

of which there are, say, m independent ones. The possible states of the system are then given by the points of an n-dimensional surface in $3N$-dimensional space, where $n = 3N - m$ is the number of "degrees of freedom" of the system. The surface itself is the phase space of the system. The reader should illustrate this with further examples.

If m_k is the mass at the kth point and if $\mathfrak{x}_k$ is the position vector of this point — variable in time, thus $\mathfrak{x}_k = \mathfrak{x}_k(t)$ — then the *kinetic energy* of the system is defined as

$$T = \sum_{k=1}^{N} \frac{m_k}{2}\,\dot{\mathfrak{x}}_k^2 = \sum_{k=1}^{N} \frac{m_k}{2}\,(\dot{x}_{1k}^2 + \dot{x}_{2k}^2 + \dot{x}_{3k}^2). \tag{14.3}$$

If we introduce the generalized coordinates $q^1, \ldots, q^n$ in place of the $\mathfrak{x}_k = \mathfrak{x}_k(q^i)$, then it follows, using $\dot{\mathfrak{x}}_k = (\partial\mathfrak{x}_k/\partial q^i)\dot{q}^i$, that

$$T = \sum_{k=1}^{N} \frac{m_k}{2}\,\frac{\partial\mathfrak{x}_k}{\partial q^i}\,\frac{\partial\mathfrak{x}_k}{\partial q^j}\,\dot{q}^i\dot{q}^j$$

which, with the notation of (14.4), can be written in the form (14.5), as follows:

$$g_{ij}(q) \underset{\text{def}}{=} \sum_{k=1}^{N} m_k\,\frac{\partial\mathfrak{x}_k}{\partial q^i}\,\frac{\partial\mathfrak{x}_k}{\partial q^j} \tag{14.4}$$

$$2T = g_{ij}(q)\dot{q}^i\dot{q}^j. \tag{14.5}$$

Since $T > 0$ holds by (14.3), unless all $\dot{q}^i$ vanish, we have the following result:

Theorem 14.1.1. *With the kinetic energy T of a dynamical system of the above type, there is associated a Riemannian metric in the phase space, given by*

$$ds^2 = 2T\,dt^2 = g_{ij}(q)\,dq^i\,dq^j, \tag{14.6}$$

where the $g_{ij}(q)$ are as in (14.4). *The element of arc given by* (14.6) *is called the kinematic element of arc of the phase space.*[1]

The physical properties of the system must all be independent of the particular choice of coordinates q^i. For (14.6), this requirement is satisfied because the g_{ij} are the components of a tensor, by (14.4). We can insure the independence of results of the coordinate system by continuing to use the notation and methods of the tensor calculus.

Exercise 14.1.2. *The Kinematic Element of Arc for the Rod in the Plane.* Above, we coordinatized this system with 3 degrees of freedom in two different ways. Assuming a unit mass at each of the two end points, *compute* the kinematic element of arc from $ds^2 = dx_1{}^2 + dx_2{}^2 + dy_1{}^2 + dy_2{}^2$, using the equation of restraint (14.1), first for the coordinates q^1, q^2, q^3 and then, using the angle $\vartheta = \bar{q}^3$, for the coordinates $q^1, q^2, \bar{q}^3$.

Exercise 14.1.3. *Prove*: For a mass point of unit mass moving on a surface, the kinematic element of arc is equal to the element of arc on the surface.

14.2 Hamilton's Principle and the Equations of Motion

A point (q^i) of phase space represents a possible state (position) of the dynamical system; a curve $q^i(t)$ is the image of a motion of the system if t is interpreted as time. We want to investigate what curves are possible, as representatives of motions of the system, if there are forces acting on the system. We shall confine ourselves here to the simple but very important case of conservative systems. For these, there exists a function $U = U(q^i)$, called a potential, such that the "generalized" forces are given by

$$P_i \underset{\text{def}}{=} -\frac{\partial U}{\partial q^i}. \tag{14.7}$$

---

[1] The qualification, "system of the above type," is essential. If one allows restraints (14.2) that depend explicitly on time (leading to "rheonomic" systems instead of the "scleronomic" ones considered here), or if one allows differential conditions — so-called "anholonomic" restraints on the coordinates — then more general metrics than Riemannian metrics are called for. The reader is referred to textbooks on mechanics, such as G. Hamel, "Theoretische Mechanik," Springer, Berlin, 1949, or E. Whittaker, "Analytical Dynamics," 4th ed., New York, 1944.

In mechanics it is proved that any motion of such a system must satisfy *Hamilton's principle*

$$\delta \int_{t_1}^{t_2} (T - U)\, dt = 0. \tag{14.8}$$

For a curve ("trajectory") $q^i(t)$ to yield a stationary value for this integral, necessary and sufficient conditions are given by the *Euler-Lagrange equations* (see Appendix II.3):

$$\frac{\partial(T-U)}{\partial q^k} - \frac{d}{dt}\frac{\partial(T-U)}{\partial \dot{q}^k} = 0. \tag{14.9}$$

Let us transform these equations by introducing the expression (14.6) for the kinematic element of arc and noting that U does not depend on $\dot{q}^k$. We have

$$\frac{\partial T}{\partial \dot{q}^k} = g_{kr}\, \dot{q}^r \underset{\text{def}}{=} p_k. \tag{14.10}$$

(These quantities p_k play an important role in Hamiltonian theory, under the name of generalized impulses.) Further,

$$\frac{\partial T}{\partial q^k} = \frac{1}{2}\frac{\partial g_{ij}}{\partial q^k}\, \dot{q}^i \dot{q}^j \tag{14.11}$$

and

$$\frac{d}{dt}\frac{\partial T}{\partial \dot{q}^k} = g_{kr}\, \ddot{q}^r + \frac{1}{2}\left(\frac{\partial g_{ki}}{\partial q^j} + \frac{\partial g_{kj}}{\partial q^i}\right)\dot{q}^i \dot{q}^j. \tag{14.12}$$

If we substitute (14.7) and (14.11), (14.12) into (14.8) and then use inner multiplication by g^{kl}, we obtain

$$\ddot{q}^l + \Gamma^l_{ij}\, \dot{q}^i \dot{q}^j = g^{lr}\, P_r \tag{14.13}$$

or, introducing the covariant derivative and the contravariant force vector $P^l \underset{\text{def}}{=} g^{lr}\, P_r$,

$$\frac{D\dot{q}^l}{dt} = P^l. \tag{14.14}$$

The trajectories of our dynamical system, under the force field given by the potential function U, are the solutions of the differential equation (14.14). The parameter t is time.

Let us discuss (14.14) for some special cases. If there are no external forces ($P^l = 0$), then the trajectories are precisely the geodesics of the kinematic metric and the time t is then a natural parameter, hence proportional to arc length. This contains a new characterization of geodesics on surfaces: If there are no external forces, then a mass point moving on a surface moves along geodesics, at constant speed.

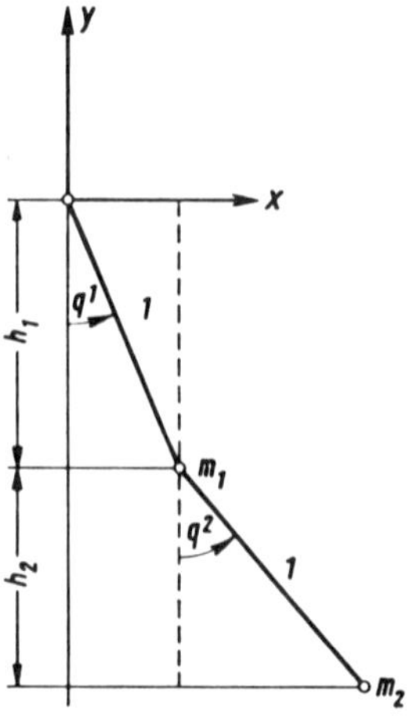

FIG. 37. The double pendulum.

This is a generalization of Newton's law of inertia which can be stated for any number of degrees of freedom as follows: A system not under the influence of any external forces moves in such a way that the trajectory in the phase space of the system is a "straightest curve," or geodesic, of that space, and this geodesic is traversed at constant speed.

If there are external forces, then (14.14) yields the following generalization of Newton's law: The acceleration (written covariantly) equals the contravariant force vector P^l, hence the motion of the system corresponds to the motion of a point with unit mass in the phase space. The external force is responsible for the deviation of the trajectory from the "straightest," i.e., geodesic, course.

From the equations of motion, we can derive the *principle of conservation of energy*: Along any trajectory we have $T + U =$ const., that is, the sum of the kinetic and the potential energy is invariant along a trajectory. Indeed, using Ricci's lemma, we obtain

$$\frac{d}{dt}(T+U) = \frac{D}{dt}\left(\frac{1}{2}g_{ik}\dot{q}^i\dot{q}^k\right) + \frac{\partial U}{\partial q^r}\dot{q}^r$$

$$= g_{rl}\dot{q}^r\left(\frac{D\dot{q}^l}{dt} - P^l\right) = 0.$$

Exercises. Obtain the equations of motion for the following systems:

14.2.1. A point of unit mass moving on a surface in the gravitational field. If g_{ik} is the fundamental tensor of the surface, then $2T\,dt^2 = g_{ik}\,dq^i\,dq^k$, and if the space is referred to a Cartesian coordinate system x_1, x_2, x_3, then $U = g \cdot x_3$, where g is to be taken as a constant (the "gravitational constant").

14.2.2. *The Double Pendulum* (see Fig. 37; both masses are unit masses, each thread is of length 1). We have $U = -g[h_1 + (h_1 + h_2)]$, $x_1 = \sin q^1$, $y_1 = -h = -\cos q^1$, also

$$x_2 = \sin q^1 + \sin q^2, \qquad y_2 = -(h_1 + h_2) = -(\cos q^1 + \cos q^2)$$

and

$$2T\,dt^2 = ds^2 = dx_1{}^2 + dy_1{}^2 + dx_2{}^2 + dy_2{}^2.$$

14.3 Jacobi's Element of Arc and the Principle of Least Action

The connection between analytical dynamics and Riemannian geometry shows up as even more intimate if, instead of using the kinematic element of arc, one uses a different metric, due to Jacobi. With a constant h (whose meaning will come out presently), let us set

$$d\tilde{s}^2 = 2(h-U)\,ds^2 = \tilde{g}_{ik}\,dq^i\,dq^k \tag{14.15}$$

where

$$\tilde{g}_{ik} = 2(h-U)\,g_{ik}. \tag{14.16}$$

Then the following result holds:

Theorem 14.3.1. *The geodesics of Jacobi's metric* (14.15) *are those trajectories of the system along which the total energy $T + U$ has the fixed value h.*

Proof. Let $q^i(t)$ be a trajectory associated with the fixed value h of the total energy $T + U$. Then, along this curve, the equations of motion (14.13) are valid, as well as $T + U = h$. The relation (mapping) between the Jacobian metric and the kinematic metric g_{ik} is conformal and therefore, by relation (*) of §13.4, the Christoffel symbols satisfy

$$\tilde{\Gamma}^l_{ij} = \Gamma^l_{ij} + \frac{1}{2(h-U)}\left(\delta^l_i(h-U)_{,j} + \delta^l_j(h-U)_{,i} - g^{lr}g_{ij}(h-U)_{,r}\right).$$

We also have

$$\frac{d\tilde{s}}{dt} = 2\sqrt{(h - U)T} = 2(h - U), \qquad \frac{d^2\tilde{s}}{dt^2} = 2(h - U)_{,r}\dot{q}^r.$$

Substituting all of this into the equations of motion, we obtain

$$\frac{d^2q^l}{d\tilde{s}^2} + \tilde{\Gamma}^l_{ij}\frac{dq^i}{d\tilde{s}}\frac{dq^j}{d\tilde{s}} = 0.$$

The trajectory is therefore a geodesic of the metric (14.16). What is more, we obtain in this way all of the geodesics of this metric, since, for any fixed h, there exists, for every direction at any given point, a trajectory with total energy h, its initial tangent being determined by $\frac{1}{2}g_{ik}\,x^i\,x^k = h - U$. We assume here that $h > U$.

Note that because $T + U = h$, we have $d\tilde{s}^2 = 4T^2\,dt^2$ along a trajectory, hence

$$A = \int d\tilde{s} = \int 2\,T\,dt.$$

The quantity $A = \int d\tilde{s}$ is called the *action* (or *effect*) along the trajectory. The fact expressed by Theorem 14.3.1 is known as *Jacobi's* formulation of the *principle of least action*. It would not be correct to say that it states that the integral $\int 2T\,dt$ must be a stationary value, because $d\tilde{s} = 2T\,dt$ holds, after all, only along the trajectories and not along neighboring curves that should be admitted for comparison in studying the variation, such neighboring curves not being trajectories (in general). In the history of mechanics, this tempting fallacy played a significant role, since the principle of least action was first enunciated by Maupertuis in the faulty formulation $\delta\int 2T\,dt = 0$. (See, for instance, E. Mach, "Die Mechanik in ihrer Entwicklung", say the 4th ed., Leipzig, 1901; in English, as "The Science of Mechanics," 5th ed., La Salle, Illinois, 1942).

Theorem 14.3.1. opens up a multitude of possibilities for applying theorems of Riemannian geometry to analytical mechanics. Let us look at some examples.

Theorem 14.3.2. *Let there be given, in the same phase space, two mechanical systems with equal kinetic energies T, T' and equal potential energies U, U'. If there are two numbers h, h' such that the trajectories with total energy h of the first system coincide with the trajectories with total energy h' of the second system,*[2] *then $U' = cU + d$, where c and d are constants. This means that the two systems are then identical, except for a change of scale.*

[2] We do not have to assume here that the trajectories are traversed with proportional speeds.

Proof. The hypotheses imply that the two associated Jacobi metrics can be related conformally and geodesically, whence Theorem 13.4.2 gives the existence of a constant d such that

$$(h - U)\, g_{ik} = d(h' - U')\, g_{ik},$$

and this implies the conclusion.

14.4 Brachystochrones

This is the name given to curves in phase space having the following minimal property: Among all curves with fixed total energy h that join two states (i.e., points of phase space) P and Q, they are traversed in the shortest possible time.

If $T + U = h$ along a curve $q^i(t)$, then

$$g_{ik}\, dq^i\, dq^k = 2(h - U)\, dt^2,$$

hence the time t taken for the traversal of an arc of the trajectory is given by

$$t = \int \sqrt{\frac{g_{ik}(dq^i/d\lambda)(dq^k/d\lambda)}{2(h - U)}}\, d\lambda. \tag{14.17}$$

Therefore, the curves that are associated with a stationary value of t are the geodesics of the Riemannian metric with fundamental tensor

$$\frac{g_{ik}}{2(h - U)}. \tag{14.18}$$

This "brachystochronic" metric, too, is conformally related to the kinematic metric and therefore also to Jacobi's metric.

Theorem 14.4.1. *If the brachystochrones with total energy h' are also the brachystochrones (not necessarily with equal or proportional speed of traversal) associated with total energy $h \neq h'$, then $U =$ const.*

Proof. It follows from Theorem 13.4.2 that there exists a constant c such that

$$\frac{g_{ik}}{2(h - U)} = c\,\frac{g_{ik}}{2(h' - U)},$$

and this implies the conclusion.

Theorem 14.4.2. *If the brachystochrones with total energy h are also trajectories (not necessarily with equal or proportional speed of traversal) with total energy h' (not necessarily $\neq h$), then $U =$ const.*

Proof. Once again, it follows that

$$\frac{g_{ik}}{2(h - U)} = 2(h' - U)\, g_{ik},$$

which implies the conclusion.

Thus it is seen that the trajectories associated with a total energy h are also brachystochrones only if $U = \text{const.}$

On the other hand, we have the following result:

Theorem 14.4.3. *The brachystochrones associated with a value h, in a system of kinetic energy T and with a potential U, are also the trajectories associated with h' of a system that has the same kinetic energy, but a potential U', where h' and U' are in general different from h and U, respectively. The converse also holds.*

Proof. If we compare (14.18) with Jacobi's metric, we see that all that is necessary is to set

$$2(h' - U') = \frac{1}{2(h - U)}.$$

Exercise 14.4.4. Find dynamical systems with two degrees of freedom such that, for suitable values of the total energy, the (a) trajectories, (b) brachystochrones are circles in the (q^1, q^2)-plane. (*Hint.* By §12.6, the geodesics of the metric

$$g_{ik} = \frac{\delta_{ik}}{(q^2)^2}$$

are circles orthogonal to the q^1-axis. Replace q^2 by $c - q^2$.)

Exercise 14.4.5. The historically earliest example of brachystochrones, and one which played an important role in the development of the calculus of variations, is given by the cycloids whose axis is parallel to the q^2-axis. *Prove* that these cycloids are brachystochrones of a mass point in the gravitational field. (*Hint.* The kinetic energy is $\frac{1}{2}[(\dot{q}^1)^2 + (\dot{q}^2)^2]$, the potential energy is $U = g \cdot q^2$ with $g = \text{const.}$ Set up the differential equation for the brachystochrones and show that the cycloids are solutions.)

14.5 The Sign of the Curvature and the Stability of Trajectories

It is to be expected that further important quantities and facts of Riemannian geometry are of interest in dynamics. Let us look at an application of the concept of curvature.

Let $q^i(t)$ be a trajectory of the system. The trajectory is said to be *stable* if all trajectories with sufficiently close initial conditions at t_0 remain close to the trajectory $q^i(t)$ for all later times t. More precisely, the definition of stability is as follows:

$q^i(t)$ is a stable trajectory if for any $\varepsilon > 0$, there exists a $\delta(\varepsilon) > 0$ such that

$$|r^i(t) - q^i(t)| < \varepsilon$$

for all t and for all trajectories $r^i(t)$ satisfying both $|r^i(t_0) - q^i(t_0)| < \delta(\varepsilon)$ and

$$|\dot{r}^i(t_0) - \dot{q}^i(t_0)| < \delta(\varepsilon).$$

To investigate conditions for stability, assume, first, the case in which $U = \text{const.}$, so that the trajectories are the same as the geodesics of the kinematic metric. We can, therefore, use the facts about the geodesic deviation developed by Jacobi and Levi-Cività (see §11.8), according to which the stability is governed by the differential equation

$$\ddot{\eta}^i + R^i{}_{jkl}\,\dot{q}^j\,\dot{q}^k\,\eta^l = 0, \tag{14.19}$$

on whose solutions the question depends in the following sense: If all solutions that correspond to small initial conditions η_0, $\dot{\eta}_0$ also remain small for later times t, then the geodesics are stable. Hence if $n = 2$, then $K > 0$ insures stability, $K \leqq 0$ instability.

If $U \neq \text{const.}$, the trajectories are still geodesics, but in terms of Jacobi's metric only. The differential equation (14.19) applies in this case with $\tilde{s}$ in place of t. Here, too, that differential equation furnishes a stability criterion. In particular, if $n = 2$ and $K > 0$ then the trajectory is stable in the sense of Jacobi's metric; note that stability here only involves comparison with trajectories with the same energy h, since only these trajectories are geodesics of the corresponding Jacobi metric.

Exercise 14.5.1. *Show* that the movement of the planets is stable in the sense of Jacobi's metric. (*Hint.* Without loss of generality, the gravitational constant and the masses of the central body and the planet can each be taken as equal to 1, so that

$$2T\,dt = \delta_{ik}\,dq^i\,dq^k, \qquad U = 1/\sqrt{(q^1)^2 + (q^2)^2}.\ )$$

§ 15. Metric Differential Geometry and Characterizations of Riemannian Geometry

15.1 The Metric and the Fundamental Tensor

When we introduced Riemannian spaces, we noted (in §11.1) that there were more general definitions possible for measuring lengths than were adopted on that occasion, and that we were taking up the special geometry that is called Riemannian for the simple reason that it is a straightforward generalization of the intrinsic geometry of surfaces. This was also Riemann's own motivation. Now, however, we wish to proceed to the study of more general possibilities for a measurement of lengths. This will subsequently enable us to observe that Riemannian geometry is distinguished, among all other metrics, not only by its particular formal simplicity, but also by certain characteristic geometric properties.

We are aiming at a definition of length that assigns to every smooth arc C of a curve in a manifold a positive number $L(C)$ as its "length" in such a way that, if C is decomposed into two subarcs C_1 and C_2, then $L(C_1) + L(C_2) = L(C)$. One way of arriving at such a definition is to set

$$L(C) = \int_{t_0}^{t_1} F(x; \dot{x})\, dt,$$

where $F(x; \dot{x})$ is a function defined for all x of the space and positive for all $\dot{x} \neq 0$. We want $L(C)$ to be independent of the parametrization and of the sense of traversal of C, which leads to the requirement

$$L(C) = \int_{t_0}^{t_1} F(x; \dot{x})\, dt = \int_{\tau_0}^{\tau_1} F\left(x; \frac{dx}{d\tau}\right) d\tau = \int_{t_0}^{t_1} F\left(x; \dot{x}\frac{dt}{d\tau}\right)\frac{d\tau}{dt}\, dt.$$

Comparing the first integral with the last, we see that this requires that

$$F(x; \lambda\xi) = |\lambda| F(x; \xi) \qquad \text{for all real } \lambda, \tag{15.1}$$

for only if (15.1) holds will the above integrals be equal for arbitrary t-intervals. We are now ready for the following definition:

Definition. *A differentiable manifold is called a metric space if there is a continuous function $F(x; \xi)$ defined for all points (x) and all tangent vectors (ξ) and having the following properties:*

$$F(x; \xi) > 0 \quad \text{for} \quad \xi \neq 0 \quad (\textit{positive definiteness in } \xi) \tag{A}$$

$$F(x; \lambda\xi) = |\lambda| F(x; \xi) \quad (\textit{positive homogeneity in } \xi \textit{ of 1st degree}). \tag{B}$$

The absolute value of

$$\int_{t_0}^{t_1} F(x; \dot{x})\, dt$$

is then defined as being the length of the arc $x(t)$, $t_0 \leqq t \leqq t_1$.

For the purposes of differential geometry, we shall usually have to assume that $F(x;\xi)$ has continuous partial derivatives of sufficiently high orders for all x and all $\xi \neq 0$.

The Riemannian spaces are special metric spaces, being associated with the metric base function (or *distance function*) $F(x;\xi) = \sqrt{g_{ik}(x)\xi^i \xi^k}$. On the basis of our experiences with Riemannian geometry, we shall try to introduce a fundamental metric tensor in the more general case also. We define

$$g_{ik}(x;\xi) \underset{\text{def}}{=} \frac{1}{2} \frac{\partial^2 F^2}{\partial \xi^i \, \partial \xi^k} \tag{15.2}$$

and we observe, to begin with, that in the Riemannian case this indeed reduces to the fundamental metric tensor $g_{ik}(x)$ (which, in the Riemannian case, is independent of ξ). Next we show that in the more general case as well, (15.2) is always a tensor. If $\bar{x}^i = \bar{x}^i(x)$ is a coordinate transformation, we apply the transformation law for contravariant vectors to ξ and obtain

$$\bar{\xi}^i = \frac{\partial \bar{x}_i}{\partial x^k} \xi^k, \qquad \text{hence} \qquad \frac{\partial \bar{\xi}_i}{\partial \xi^k} = \frac{\partial \bar{x}^i}{\partial x^k},$$

and we must also have $\bar{F}(\bar{x};\bar{\xi}) = F(x;\xi)$ for invariance of "length" under the coordinate transformation. Hence

$$2g_{ik}(x;\xi) = \frac{\partial^2 F^2}{\partial \xi^i \, \partial \xi^k} = \frac{\partial}{\partial \xi^i} \frac{\partial F^2}{\partial \xi^k} = \frac{\partial}{\partial \xi^i} \frac{\partial \bar{F}^2}{\partial \bar{\xi}^r} \frac{\partial \bar{x}^r}{\partial x^k} = \frac{\partial^2 \bar{F}^2}{\partial \bar{\xi}^r \, \partial \bar{\xi}^l} \frac{\partial \bar{x}^r}{\partial x^k} \frac{\partial \bar{x}^l}{\partial x^i}$$

$$= 2\bar{g}_{rl}(\bar{x};\bar{\xi}) \frac{\partial \bar{x}^r}{\partial x^i} \frac{\partial \bar{x}^l}{\partial x^k},$$

and this is precisely the tensorial transformation law for a doubly covariant tensor.

We recall, at this point, Euler's identity for homogeneous functions: If $f(\xi)$ is homogeneous of degree n, i.e., if $f(\lambda \cdot \xi) = \lambda^n f(\xi)$, then

$$\frac{\partial f}{\partial \xi^i} \xi^i = n \cdot f(\xi).$$

(See any textbook on the calculus or advanced calculus, e.g., R. Courant, "Differential and Integral Calculus," Vol. II, p. 109, New York, 1940.) This shows that differentiation with respect to ξ lowers the degree of homogeneity by 1. If we apply Euler's relation to the function $g_{ik} = \frac{1}{2}\partial^2 F^2 / \partial \xi^i \, \partial \xi^k$, which is homogeneous in ξ of degree 0, and if we also make use of the equality of the mixed partials, we obtain the important relations

$$\frac{\partial g_{ik}}{\partial \xi_j} \xi^j = \frac{\partial g_{ij}}{\partial \xi^k} \xi^j = 0, \qquad F^2 = g_{ik} \xi^i \xi^k. \tag{15.3}$$

The differential geometry of metric spaces is of particular interest in a special case that has been studied intensively, beginning with P. Finsler (1918), and which is defined as follows:

A metric space is called a Finsler space if

$$g_{ik}(x; \xi)\eta^i \eta^k > 0 \qquad \textit{for} \qquad \xi \neq 0, \quad \eta \neq 0, \tag{C}$$

that is, if the quadratic form associated with g_{ik} *is positive-definite.*

It is clear that any Riemannian space is also a Finsler space. Finsler spaces share with Riemannian spaces the property that their geodesic curves are uniquely determined as the shortest paths joining any two sufficiently neighboring points. It follows from (C) that $\det(g_{ik}) \neq 0$, so that there exists a symmetric doubly contravariant tensor $g^{ik}(x; \xi)$ such that

$$g^{ij} g_{jk} = \delta^i_k. \tag{15.4}$$

15.2 The Pointwise Minkowski Metric and the Indicatrix

In the tangent space at a fixed point (x_0) of a metric space, let a metric be defined as follows:

$$F(\xi) = F(x; \xi), \qquad \text{where} \qquad \begin{cases} \text{(A)} & F(\xi) > 0 \quad \text{for} \quad \xi \neq 0 \\ \text{(B)} & F(\lambda\xi) = |\lambda| F(\xi). \end{cases} \tag{15.5}$$

In the general geometry that corresponds to this metric, the place of the Euclidean unit sphere is taken by the hypersurface (see Fig. 38)

$$F(\xi) = 1. \tag{15.6}$$

This hypersurface is called the *indicatrix of the Minkowski metric* (15.5). It is a closed, continuous surface with center (of symmetry) 0, and it is intersected exactly once by every ray emanating from 0, which can be seen as follows: If ξ^i is any nonzero vector, then $F(\xi) = \lambda > 0$, hence $F(\xi/\lambda) = 1$ by (15.5 B). Therefore ξ/λ is the uniquely determined vector in the direction of ξ whose end point lies on the indicatrix. Further, since $F(-\xi) = F(\xi)$, the origin is the center of the indicatrix; that is to say, if ξ is a point on the indicatrix, then so is $-\xi$.

Conversely, let J be any closed hypersurface with center at 0 and such that every ray emanating from 0 intersects J in exactly one point; then there is a Minkowski metric whose indicatrix is J. For, if ξ is any vector, then the hypotheses imply the existence of precisely one positive multiple $\mu \cdot \xi$ such that the end point of $\mu \cdot \xi$ lies on the hypersurface J. If we then define $F(\xi) = 1/\mu$, we have a metric with the properties (15.5).

Thus a Minkowski metric is completely known if its indicatrix is given. The metric is Euclidean if (and only if) the indicatrix is an ellipsoid with center at 0, given by $g_{ik}\,\xi^i\,\xi^k = 1$. Just as we can say that Riemannian spaces are pointwise Euclidean, so we can say that metric spaces are pointwise Minkowskian.

We shall now investigate a procedure for associating with a given Minkowski metric in a vector space a uniquely determined Euclidean metric in the same space; this then allows the assignment of a Riemannian metric

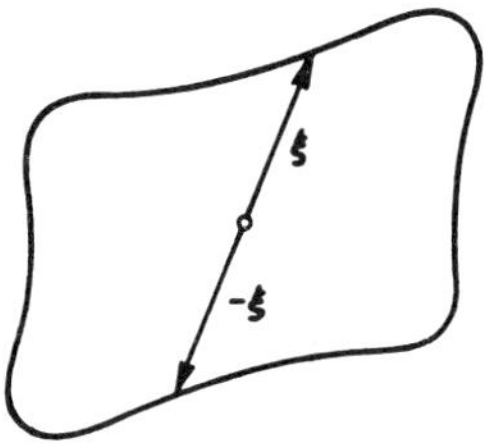

FIG. 38. Indicatrix.

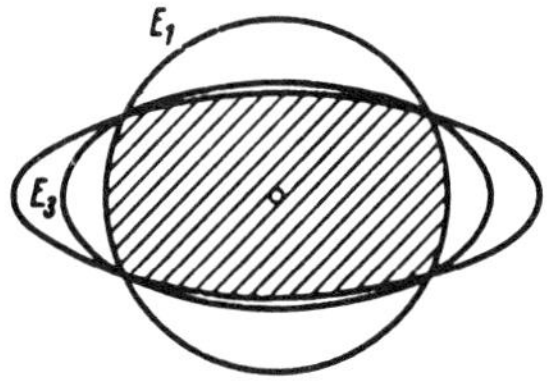

FIG. 39. Uniqueness of minimal ellipsoid.

to a given metric space. To achieve this, we shall assign to the Minkowski indicatrix of the tangent space at any given point an ellipsoid, to serve as the indicatrix at the point of the associated Riemannian geometry. This is done by means of an elementary lemma:

Lemma. (K. Löwner). *Among all ellipsoids with center at the origin 0 of a vector space and circumscribed about a given hypersurface J, there is a unique ellipsoid of smallest volume.*

("Circumscribed" means that the points of J are inside and on the ellipsoid.)

Proof. (a) *Existence* of the minimal ellipsoid. Since the function F is continuous, the surface $F(\xi) = 1$ is bounded. Therefore there do exist ellipsoids that contain J. Let V be the infimum of the volumes of (the regions bounded by) all such ellipsoids. We have $V > 0$ because all circumscribed ellipsoids contain the surface J, inside which there is contained a sphere of positive radius, again for reasons of continuity. There exists a sequence of these ellipsoids such that their volumes converge to V. From this sequence we can select, by a well-known principle of analysis, a subsequence that actually converges to an ellipsoid (say by selecting first a subsequence for which the vectors in the direction of the largest principal axis converge, etc.).

(b) *Uniqueness* of the minimal ellipsoid. Let us assume there are two ellipsoids of minimal volume V, each circumscribed about J. We choose a coordinate system in the vector space such that the two ellipsoids are given by (see Fig. 39)

$$E_1: \qquad f_1(\xi) \underset{\text{def}}{=} a(\xi_1^2 + \xi_2^2 + \ldots + \xi_n^2) = 1$$

$$E_2: \qquad f_2(\xi) \underset{\text{def}}{=} g_1\, \xi_1^2 + g_2\, \xi_2^2 + \ldots + g_n\, \xi_n^2 = 1$$

(this can be done by a "principal axis transformation"). Here a and the g_i are positive and, since the two ellipsoids have the same volume, we have $a^n = \Pi g_i$. (The volume of the ellipsoid E_2, whose semiaxes are of lengths $1/\sqrt{g_i}$, equals the volume of the unit ball multiplied by the product $1/\sqrt{g_1 \ldots g_n}$ of the lengths of the semiaxes.) For any point ξ of the surface J, we have $f_i(\xi) \leqq 1$ for $i = 1, 2$ since all of J is inside or on each of the two ellipsoids. We now consider the ellipsoid

$$E_3: \qquad f_3(\xi) \underset{\text{def}}{=} \tfrac{1}{2}(f_1(\xi) + f_2(\xi)) = 1.$$

Since $f_i(\xi) \leqq 1$ for $i = 1, 2$ implies that $f_3(\xi) \leqq 1$, it follows that E_3, too, is circumscribed about J. We shall now show that the volume of E_3 is less than V unless E_1 and E_2 coincide, which will establish uniqueness (since V is the smallest possible volume). Indeed, the semiaxes of E_3 are of lengths $\sqrt{2}/\sqrt{a + g_i}$, and, by the theorem on the arithmetic and the geometric mean, we have $a + g_i \geqq 2\sqrt{a \cdot g_i}$ with equality holding only for $a = g_i$. Denoting the volume of the unit ball by W, we therefore have for the volume of E_3:

$$W \cdot \Pi \frac{1}{\sqrt{(a+g_i)/2}} \leqslant W\, \Pi \frac{1}{\sqrt[4]{a}\,\sqrt[4]{g_i}}$$

$$= \sqrt{W \sqrt{a^{-n}}} \sqrt{W\, \Pi \frac{1}{\sqrt{g_i}}} = \sqrt{V} \cdot \sqrt{V} = V,$$

with equality holding only if all $g_i = a$, that is, only if E_1 and E_2 coincide. This completes the proof of Löwner's lemma.

We have thus obtained, in our given vector space, a Euclidean metric whose indicatrix is the minimal ellipsoid circumscribed about the indicatrix of the Minkowski metric at the given point. We speak of this as the Euclidean metric associated with the Minkowski metric at the point. If we carry out this construction in every tangent space (that is, at every point) of a metric space, then we obtain for the underlying space a Riemannian metric associated with the given metric (15.1). It is not a foregone conclusion that the fundamental metric tensor $g_{ik}(x)$ of this associated Riemannian metric is differentiable, even if the fundamental tensor $g_{ik}(x; \xi)$ of the given metric is differentiable; however, in the problems that we shall take up we shall be able to furnish direct verifications of differentiability.

If the given metric is itself Riemannian, then the associated Riemannian metric is obviously the same as the given metric.

Exercise 15.2.1. With the plane as the underlying manifold, consider the (general) metrics with the following indicatrices: (a) A square with center at 0 and sides parallel to the axes; (b) A square with its vertices on the axes; (c) A regular hexagon. For the circumference of the indicatrix (measured in the Minkowski metric which it generates), *prove* that its length equals 8 in cases (a) and (b), and equals 6 in case (c). (More generally, it can be shown that this circumference, which corresponds to the Euclidean 2π, lies between 6 and 8 for any convex indicatrix.) For the first two cases, *prove* also that the corresponding distance functions are

$$\text{(a)} \quad F(\xi) = \max(|\xi^1|, |\xi^2|); \qquad \text{(b)} \quad F(\xi) = |\xi^1| + |\xi^2|.$$

Exercise 15.2.2. *Prove*: If $f(\xi)$ is the distance function of a Minkowski metric and if $\overset{i}{B}_k(x)$ is any tensor field such that $\det(\overset{i}{B}_k(x)) \neq 0$ for all x, then the function

$$F(x; \xi) = f(B^i{}_k(x)\xi^k)$$

is the metric base function of a metric space.

Exercise 15.2.3. *Prove*: The requirement $g_{ik}(\xi)\eta^i \eta^k > 0$ implies convexity of the indicatrix.

15.3 Helmholtz' Characterization of Riemannian Spaces in Terms of the Existence of Pointwise Rotations

We turn now to the problem of finding intrinsic properties of Riemannian spaces that distinguish these spaces among all other metric spaces, and we begin with a characterization found in 1868 by Helmholtz.

Let $y^i = y^i(x)$ be a mapping of a metric space onto itself under which the point x_0 remains fixed. For the contravariant vectors ξ in the tangent space at (x_0), this induces a mapping

$$\eta^i = \frac{\partial y^i(x_0)}{\partial x^k} \xi^k = \overset{i}{A}_k \xi^k, \tag{15.7}$$

which is homogeneous linear. If this mapping (15.7) is such that it leaves the lengths of vectors invariant, that is if $F(x_0; \eta) = F(x_0; \xi)$ holds for any ξ, η related by (15.7), then we may call it a (*pointwise*) *rotation* about (x_0). The physical background which motivated Helmholtz is as follows. A pointwise rotation gives an approximately isometric mapping onto itself of a sufficiently small neighborhood of (x_0), and it will be the more closely isometric the smaller the portion of space near (x_0) that we are confined to. For by Taylor's formula,

$$y^i(x_0 + \varepsilon\, \xi^i) = x_0^i + \varepsilon\eta^i + \varepsilon^2(\ldots),$$

so that to within second-order terms in ε, we have

$$F(x_0; \Delta y) = F(x_0; \varepsilon\eta^i) = F(x_0; \varepsilon\xi^i) = F(x_0; \Delta x), \qquad \Delta x = \varepsilon\xi^i,$$

and this gives an approximation for the distance from $x_0 = y(x_0)$ to $x_0 + \Delta y = y(x_0 + \Delta x)$. Helmholtz expressed this fact in the following graphic manner: A mapping with this property moves sufficiently small bodies "rigidly," that is, without changing their measurements. For our purposes, all that counts is the (exactly) isometric mapping (15.7) of the tangent space onto itself. We shall show that the existence of sufficiently many pointwise rotations is enough to insure the Riemannian character of the space:

Theorem of Helmholtz. *Assume that at every point of a metric space, there is for every pair of directions in the tangent space an isometry of the tangent space that maps the first direction onto the second. Then the Minkowski metric is Euclidean at every point.* (*Expressed differently: If about every point of a metric space, there are sufficiently many pointwise rotations that any direction at the point can be mapped onto any other direction by one of these rotations, then the space is Riemannian. Hence only in Riemannian spaces can sufficiently small rigid bodies be rotated freely.*)

Proof. Assign two directions at a point, represented by their unit vectors ξ and η, so that $F(\xi) = F(\eta) = 1$. By assumption, there exists a homogeneous linear mapping A of the tangent space onto itself (see Fig. 40) that, being by assumption an isometry, maps the indicatrix J onto itself and maps ξ onto η. Since A maps the inside of J onto itself, the mapping A is volume-preserving. Therefore it maps the minimal ellipsoid onto an ellipsoid that encloses an equal volume and is itself circumscribed about the indicatrix J; thus the image ellipsoid coincides with its pre-image, on account of the uniqueness of the minimal ellipsoid. Hence A not only maps J onto itself but also maps the minimal ellipsoid onto itself. Let ξ be a point common to the two surfaces (such a point must exist, since the ellipsoid would not be minimal otherwise). We shall prove that for any other direction as well, the ray from the center in that direction intersects the two surfaces at a common point. In fact, there is a linear isometry A that maps ξ onto a vector in that other direction, and the end point of the image vector must lie on J (which is mapped onto itself by A) as well as on the ellipsoid (for the analogous reason). Hence the surface J coincides with its minimal ellipsoid, and this implies the conclusion.

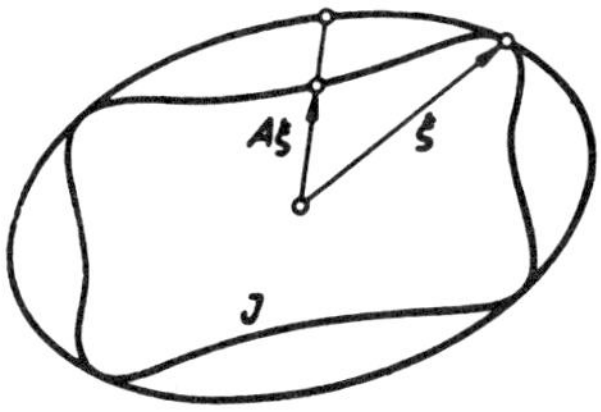

FIG. 40. Theorem of Helmholtz.

The converse fact, namely, that at any point of a Riemannian space there exist enough pointwise rotations to map any given direction onto any other, is easily established. By a suitable choice of coordinates we can always arrange that $g_{ik} = \delta_{ik}$, and then the ordinary rotations of the tangent space are pointwise rotations for the metric.

Helmholtz published this theorem together with a theorem (our 12.4.2) characterizing spaces of constant curvature. These two theorems together yield the following statement:

If sufficiently small rigid bodies can be rotated freely (that is, if free pointwise rotation is possible), then the space is Riemannian. If there is free mobility of rigid bodies of any finite extension, then the space is Riemannian of constant curvature.

15.4 Weyl's Characterization of Riemannian Spaces in Terms of the Existence of Affine Connections

We already know of the importance for Riemannian geometry of the existence of an affine connection compatible with the metric (Theorem 11.2.1), a connection which is uniquely determined and whose components are the Christoffel symbols Γ^i_{jk}. It seems natural to ask, in what metric spaces does there exist an affine connection compatible with the metric, that is, such that lengths of vectors remain unchanged under the parallel displacement given by the connection? To begin with, if such an affine connection exists, then its associated parallel displacement induces linear mappings onto each other of the Minkowski indicatrices at different points. Therefore, all of these indicatrices are images under linear mapping of any (fixed) one of them, and accordingly the following holds for the metric base function of the pointwise Minkowski metrics:

$$F(x;\xi) = f(B^i_k(x)\xi^k), \tag{15.8}$$

where f is the distance function of the Minkowski metric in a fixed one of the tangent spaces. Following Weyl, we shall call the metrics (15.8) *metrically homogeneous*, because at every point there is a coordinate system in which $B^i_k(x_0) = \delta^i_k$. The metrically homogeneous spaces include all Riemannian spaces, as well as all (globally) Minkowski spaces. In the latter we even have $B^i_k(x) = \delta^i_k$ for all points (x) if a linear coordinate system is used.

In every Riemannian space and in every Minkowski space, there exists a compatible affine connection; in the latter case, all of its components are 0 if the coordinate system is Cartesian. Weyl observed that not all metrically homogeneous spaces can be equipped with a compatible affine connection, and he even proved the following[1]:

Weyl's Theorem. *Let $f(\xi)$ be the distance function of a Minkowski metric. Assume that for every nondegenerate tensor field $B^i_k(x)$, the metric*

$$F(x;\xi) = f(B^i_k(x)\xi^k) \tag{15.9}$$

has a compatible affine connection. Then f^2 is a positive-define quadratic form; hence, the metric (15.9) *is Riemannian.*

Proof: We may assume, without loss of generality, that the associated Euclidean metric of f has the fundamental metric tensor δ_{ik}, hence that, in suitable coordinates, the minimal ellipsoid of the indicatrix $f = 1$ has the equation of the unit sphere (see Fig. 41). Then for the Riemannian metric associated with (15.9), we have

[1] Weyl first posed the problem in his commentary (1918) to Riemann's inaugural address. A solution, sufficiently general to include even the case of indefinite metrics, is to be found in Weyl's monograph "Die mathematische Analyse des Raumproblems," Berlin, 1923.

$$G(x;\xi) = \sqrt{\delta_{ik}\, B^i_r\, B^k_l\, \xi^r\, \xi^l}. \tag{15.10}$$

An affine connection compatible with the metric (15.9) induces a mapping of the indicatrix at a point (x_0) onto the indicatrix at any other point (x); this mapping takes the minimal ellipsoid at (x_0) onto that at (x), since the minimal ellipsoid is uniquely determined. Hence the affine connection is compatible not only with the metric F, but also with the associated Riemannian metric G. The components of this connection are therefore the Christoffel symbols of the metric G and this proves that there exists at most one affine connection compatible with the metric F.

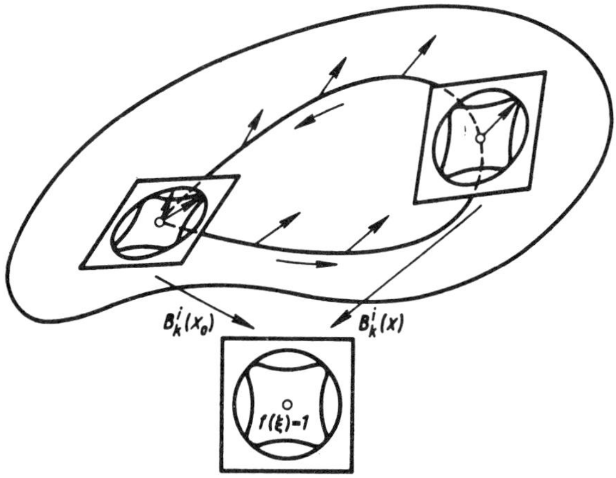

FIG. 41. Proof of Weyl's theorem.

Next we exploit the freedom still at our disposal in the choice of the tensor field $B^i_k(x)$. We choose this field in such a way that it yields a metric (15.10) that is of constant nonzero curvature K; then the holonomy group at a point is capable of mapping any given direction in the tangent space onto any other direction. The group maps the indicatrix onto itself (because the connection is compatible with the metric F), and it maps its minimal ellipsoid onto itself. From this we conclude, exactly as in the proof of Helmholtz' theorem, that the indicatrix is an ellipsoid, hence the metric Riemannian.

Thus it is seen that it is essentially only in Riemannian spaces that compatible affine connections exist. Note, however, that Weyl's theorem shows only this much: The Riemannian spaces are the only ones which have compatible affine connections for all possible choices of $B^i_k(x)$; for individual tensor fields $B^i_k(x)$, there may very well exist compatible affine connections even in non-Riemannian cases — as shown, for instance, by the example $B^i_k(x) = \delta^i_k$ in any (globally) Minkowskian space. At any rate, the proof of Weyl's theorem also yields the following:

Corollary. *There exists at most one affine connection compatible with the given metric.*

Exercise 15.4.1. *Prove*: In every metrically homogeneous space there exists a linear connection compatible with the metric (which however is, in general, not symmetric, by Weyl's theorem).

[*Hint.* If $F(x;\xi) = f(B^i_k(x)\xi^k)$, then the linear mappings $A^i_k(x)$ that are the inverses of the B^i_k (thus: $A^i_r B^r_k = \delta^i_k$) are isometric linear mappings from the tangent space at (x_0) to the tangent spaces at the other points (x) of the manifold. For every vector ξ^i_0, the vector

$$\xi^i(x) = A^i_k(x)\xi^i_0$$

is then a vector of the same length as ξ^i_0 itself. For a field of this sort, we have

$$d\xi^i = A^i_{k,j}\,\xi^k_0\,dx^j = A^i_{k,j}\,B^k_r\,\xi^r\,dx^j,$$

and this is a length-preserving parallel displacement whose coefficients are $\Gamma^i_{jr} = -A^i_{k,j}B^k_r$.]

15.5 *Riemannian Spaces as the Only Pointwise Isotropic Spaces*

In the last two subsections we proved two characterizations of Riemannian geometry that require certain linearity properties, namely, either invariance of the metric under sufficiently many linear mappings of the tangent space onto itself or invariance of the metric relative to an affine connection. To these we shall now add a proof of the following characterization: The Riemannian spaces are the only ones for which, at every point, the directions of the tangent space are all "equally good," meaning that the metric alone does not in any way single out one direction from among all directions at the point. This property deals only with the tangent spaces and may be called the pointwise isotropic property. In order to formulate it precisely, we must define, for a fixed tangent space in a metric space, the concept of a function of direction that depends on the metric alone: Such a function is defined to be a functional of the metric in the tangent space and also a function of direction, thus:

$$I = I(\xi^i; F)$$

with

$$I(\lambda\xi^i; F) = I(\xi^i; F) \qquad \text{for} \qquad \lambda > 0.$$

This last requirement is necessary so as to make I depend only on the direction of ξ and not on ξ itself. Hence any explicit expression for I can only contain F and ξ and quantities derived from these, such as, for instance, the coefficients of the equation for the minimal ellipsoid of the indicatrix J.

Theorem 15.5.1. *A metric space is Riemannian if and only if it is pointwise isotropic, that is, if and only if every function of direction that depends only on the metric is constant in the tangent space* $T(P)$.

Proof. (a) *Necessity*: In a Euclidean space, every direction ξ^i can be mapped without a change of the metric onto any other direction η^i, namely, by a rotation. If we introduce new coordinates such that $\bar{\eta}^i = \xi^i$, then

$$I(\eta^i; F) = I(\bar{\eta}^i; \bar{F}) = I(\xi^i; F),$$

because the unit sphere $\bar{J}$ has the same equation as J.

(b) *Sufficiency*: Let the minimal circumscribed ellipsoid of the indicatrix have the equation $a_{ik}\, \xi^i\, \xi^k = 1$. Then

$$I(\xi; F) = \frac{F^2(\xi)}{a_{ik}\, \xi^i\, \xi^k}$$

is a functional that satisfies the condition of being a function of direction and metric only, since the metric base function F and the minimal ellipsoid are determined solely by the metric. Hence, the functional $I(\xi; F)$ must have the same value for all directions. Since $I = 1$ at a point common to J and to the ellipsoid, it follows that $I = 1$ for all directions and this implies that the indicatrix coincides with its minimal ellipsoid.

The characterization just proved can be given the following physical interpretation: At any point, there do not exist any objective characteristics, determinable by measurements alone, that allow the singling out of one direction among all others.

A counterpart to this, in a certain sense, is the characterization of the spaces of constant curvature in terms of an infinitesimal isotropic property, as given by Schur's Theorem 12.2.2: Assuming that every function of direction that depends only on the metric and its derivatives with respect to x is constant for all planar directions at a point, it follows that $K = \text{const.}$ in the whole space. For, the Gaussian curvature is a functional that satisfies the assumptions. We speak of an infinitesimal isotropic property because not only the metric, but also its derivatives with respect to x may enter.

15.6 Geodesic Curves

We return now to the study of general metric spaces. We shall, however, assume a Finsler space in what follows so that the existence of the tensor g^{ij} is guaranteed.

Theorem 15.6.1. *The geodesics of a Finsler metric* $g_{ik}(x; \xi)$, *that is the extremal curves of the variational problem*

$$\delta \int ds = \delta \int_{t_0}^{t_1} F(x; \dot{x})\, dt = \delta \int_{t_0}^{t_1} \sqrt{g_{ik}(x; \dot{x}) \dot{x}^i \dot{x}^k}\, dt = 0, \tag{15.11}$$

are the solutions of the differential equations

$$\frac{d^2 x^i}{ds^2} + \Gamma^i_{jk}(x; \dot{x}) \frac{d\dot{x}^j}{ds} \frac{d\dot{x}^k}{ds} = 0, \tag{15.12}$$

where the Christoffel symbols

$$\Gamma^i_{jk} = \frac{g^{ir}}{2} \left(\frac{\partial g_{rj}}{\partial x^k} - \frac{\partial g_{jk}}{\partial x^r} + \frac{\partial g_{rk}}{\partial x^j} \right) \tag{15.13}$$

are formally defined exactly as in the Riemannian case, but may here depend also on the direction $(\dot{x}^i)$.

Proof. We have to show that (15.12) are the Euler-Lagrange equations of the variational problem (15.11). A computation yields

$$\frac{\partial F}{\partial x^i} = \frac{1}{2} \frac{\partial g_{jk}}{\partial x^i} \dot{x}^j \dot{x}^k / F$$

and

$$\frac{\partial F}{\partial \dot{x}^i} = g_{ij} \dot{x}^j / F,$$

where we have made use of Euler's relation (15.3). Using that same relation, and choosing $t = s$ so that the radical equals 1, we obtain

$$\frac{d}{dt} \frac{\partial F}{\partial \dot{x}^i} = \frac{dg_{ij}}{dt} \dot{x}^j + g_{ij} \ddot{x}^j = \frac{\partial g_{ij}}{\partial x^r} \dot{x}^r \dot{x}^j + g_{ij} \ddot{x}^j.$$

Substituting these results into the Euler-Lagrange equations (see Appendix II.3) and using inner multiplication by g^{li}, we obtain (15.12).

Exercise 15.6.2. *Find* the transformation law of the Christoffel symbols (15.13). [*Hint.* Differentiation of the transformtion law for the fundamental metric tensor, in § 15.1, yields

$$\frac{\partial g_{ik}}{\partial x^j} = \frac{\partial \bar{g}_{rl}}{\partial \bar{x}^m}\frac{\partial \bar{x}^m}{\partial x^j}\frac{\partial \bar{x}^r}{\partial x^i}\frac{\partial \bar{x}^l}{\partial x^k} + 2\bar{g}_{rl}\frac{\partial^2 \bar{x}^r}{\partial x^i \partial x^j}\frac{\partial \bar{x}^l}{\partial x^k} + \frac{\partial \bar{g}_{rl}}{\partial \bar{\xi}^m}\frac{\partial^2 \bar{x}^m}{\partial x^t \partial x^j}\xi^t\frac{\partial \bar{x}^r}{\partial x^i}\frac{\partial \bar{x}^l}{\partial x^k},$$

where the relation $\bar{\xi}^m = (\partial \bar{x}^m/\partial x^t)\xi^t$ was used in the last term. This last term makes the transformation law for the Christoffel symbols more complicated here than in Riemannian geometry.]

Exercise 15.6.3. Use Exercise 15.6.2 to show the following: If $\eta^i(x(t))$ is a vector field along a curve $x^i(t)$, then

$$\frac{d\eta^i}{dt} + \Gamma^{\ i}_{jk}\eta^j\dot{x}^k$$

is *not*, in general, a vector. (The transformation law for the Christoffel symbols is responsible for this. Thus we cannot use this expression as a covariant derivative in this case.)

15.7 Systems of Paths. Nonlinear Connections

Riemannian spaces are special cases of spaces with an affine connection and the compatible affine connection of a Riemannian space is closely related, through the Christoffel symbols, to the differential equation of the geodesics. In an analogous fashion, Finsler spaces can be considered as being special cases of "spaces of paths," and, here too, we use this fact to arrive at a covariant differentiation process.

Let

$$\ddot{x}^i + 2G^i(x;\dot{x}) = 0; \qquad G^i(x;\lambda\dot{x}) = \lambda^2 G^i(x;\dot{x}) \tag{15.14}$$

be the equations of a system of paths. The requirement of second-degree homogeneity, in (15.14), implies that at every point, there exists, for any given direction, exactly one path of the system that passes through the point in the given direction. (*Exercise*: Prove this.) Clearly, the geodesics of a Finsler space constitute an example of such a system of paths. Another example is furnished by the auto-parallels of an affine connection. In the case of a Finsler geometry, we have $2G^i = \Gamma^{\ i}_{jk}\dot{x}^j\dot{x}^k$; in the case of an affine connection, we have

$$2G^i = \Lambda^i_{jk}\dot{x}^j\dot{x}^k.$$

The differential equation of the parallel displacement associated with the affine connection is

$$0 = \dot{\eta}^i + G^i_k(x;\eta)\dot{x}^k = \dot{\eta}^i + \Lambda^i_{jk}(x)\eta^j\,\dot{x}^k$$

where

$$G^i_k(x;\dot{x}) \underset{\text{def}}{=} \frac{\partial G^i}{\partial \dot{x}^k}. \tag{15.15}$$

In this way, therefore, we can obtain the parallel displacement from the differential equations of the system of paths. Next we shall show that, in exactly the same way, we can obtain a type of parallel displacement from an arbitrary system of paths.

Theorem 15.7.1. *Let a system of paths be given by* (15.14). *Then there can be associated with it a covariant differentiation for vector fields* $\eta^i(t)$ *along arbitrary curves* $x^i(t)$: *Specifically,*

$$\frac{D\eta^i}{dt} \underset{\text{def}}{=} \dot{\eta}^i + G^i_k(x;\eta)\dot{x}^k \tag{15.16}$$

is also a vector field. A vector field $\eta^i(t)$ *for which* $D\eta^i/dt = 0$ *can thus be called "parallel" along* $x^i(t)$.

Proof. We need first the transformation laws for the quantities G^i and G^i_k. Since $\dot{\bar{x}}^i = (\partial \bar{x}^l/\partial x^r)\dot{x}^r$, it follows that

$$\ddot{\bar{x}}^i = \frac{\partial \bar{x}^i}{\partial x^r}\ddot{x}^r + \frac{\partial^2 \bar{x}^i}{\partial x^r\,\partial x^l}\dot{x}^r\,\dot{x}^l \qquad \text{and} \qquad \frac{\partial \dot{\bar{x}}^i}{\partial \dot{x}^r} = \frac{\partial \bar{x}^i}{\partial x^r},$$

and from $\ddot{x}^i = -2G^i$, $\ddot{\bar{x}}^i = -2\bar{G}^i$ we obtain

$$\bar{G}^i = \frac{\partial \bar{x}^i}{\partial x^r}G^r - \frac{1}{2}\frac{\partial^2 \bar{x}^i}{\partial x^r\,\partial x^l}\dot{x}^r\,\dot{x}^l. \tag{15.17}$$

Substituting all of this, we obtain from (15.16) that for $\bar{\eta}^i = (\partial \bar{x}^i/\partial x^r)\eta^r$,

$$\frac{D\bar{\eta}^i}{dt} = \frac{\partial \bar{x}^i}{\partial x^r}\frac{D\eta^r}{dt},$$

and thus the vector character of (15.16) is established and our definitions are justified.

The following theorem shows that in these definitions we have found a reasonable generalization, for Finsler spaces, of Levi-Cività's notion of parallel displacement:

Theorem 15.7.2. *Under a parallel displacement in accordance with*

$$\dot{\xi}^i + G^i_r(x;\xi)\dot{x}^r = 0$$

in a Finsler space, the length of a vector ξ^i *remains invariant.*

Proof. Let us first compute the G^i_r for this case. We have

$$G^i = \tfrac{1}{2}\Gamma^{i}_{jk}(x;\xi)\xi^j\xi^k,$$

and differentiation with respect to ξ^r yields

$$G^i_r(x;\xi) = \Gamma^{i}_{rj}\xi^j + \frac{1}{4}\frac{\partial g^{il}}{\partial \xi^r}\left(\frac{\partial g_{lj}}{\partial x^k} - \frac{\partial g_{jk}}{\partial x^l} + \frac{\partial g_{lk}}{\partial x^j}\right)\xi^j\xi^k,$$

because the expressions that result when the derivatives of the $g_{..}$ are differentiated with respect to ξ, vanish on account of Euler's relation (15.3). From $g^{il}g_{ls} = \delta^i_s$, we obtain

$$\frac{\partial g^{il}}{\partial \xi^r}g_{ls} + g^{il}\frac{\partial g_{ls}}{\partial \xi^r} = 0$$

or

$$\frac{\partial g^{il}}{\partial \xi^r} = -\,g^{sl}g^{im}\frac{\partial g_{ms}}{\partial \xi^r}.$$

Substitution of this yields

$$G^i_r = \Gamma^{i}_{rj}\xi^j - \frac{1}{2}g^{im}\Gamma^{s}_{jk}\frac{\partial g_{ms}}{\partial \xi^r}\xi^j\xi^k, \tag{15.18}$$

where all functions are to be evaluated at $(x;\xi)$.

Now, to the proof proper: We must prove that if ξ is a parallel vector field, i.e., if $d\xi^i/dt = -\,G^i_r(x;\xi)\dot{x}^r$, then

$$\frac{d}{dt}(g_{ik}(x;\xi)\xi^i\xi^k) = 0.$$

Now by (15.18), we have

$$\frac{d}{dt}(g_{ik}\xi^i\xi^k) = \left(\frac{\partial g_{ik}}{\partial x^l} - g_{rk}\Gamma^{r}_{il} - g_{ri}\Gamma^{r}_{kl}\right)\xi^i\xi^k\dot{x}^l$$

$$-\,2g_{ik}\xi^k\dot{x}^r\left(-\frac{1}{2}g^{im}\Gamma^{s}_{jk}\frac{\partial g_{ms}}{\partial \xi^t}\xi^j\xi^t\right).$$

The first term on the right vanishes, by the definition of the Christoffel symbols. As to the second term, since $g_{ik} g^{im} = \delta_k^m$, this term contains the factor $(\partial g_{ms}/\partial \xi^t)\xi^m$, which is equal to 0 by Euler's relation (15.3).

This new parallel displacement has an important property in common with Levi-Cività's parallelism, namely the property of being length-preserving. However, there is an important formal difference: The coefficients G_r^i are not linear in ξ in the present case, but only homogeneous in ξ of the first degree. This is unavoidable, since we know that length-preserving affine connections do not exist, in general, in Finsler spaces. In many proofs, this absence of linearity is no hindrance, as will be illustrated by an example in the next subsection.

For a historical account and for the further development of Finsler spaces, the reader is referred to the monograph by H. Rund, "The Differential Geometry of Finsler Spaces," Berlin-Göttingen-Heidelberg, 1959 (Springer "Grundlehren" series, Vol. 101). Here we mention only that connections in Finsler spaces were introduced in 1925 by Berwald, Synge, and Taylor, and in 1934 by É. Cartan. Connections in the geometry of paths were introduced in 1929 by Knebelman. The nonlinear connection discussed in this subsection, which is distinguished by the length-preserving character of its parallel displacement, was discovered in 1953 by W. Barthel.

Exercise 15.7.3. A similarity mapping ($d\tilde{s} = c \cdot ds$, $c =$ const.) between two Finsler spaces, referred to the same coordinates, satisfies $\tilde{g}_{ik} = c^2 g_{ik}$. *Prove* that the two spaces are equipped with the same connection G_k^i.

15.8 Determination of a Finsler Metric by a System of Paths

With every Finsler metric we can associate the system of paths consisting of its geodesics. The converse question is answered by the following theorem.

Theorem 15.8.1. *The paths of a system* $\ddot{x}^i + 2H^i(x; \dot{x}) = 0$ *are the geodesics of a Finsler metric* $F(x; \xi)$ *if and only if* F *is invariant under the parallel displacement*

$$\frac{d\xi^i}{dt} + H_r^i(x; \xi)\frac{dx^r}{dt} = 0$$

associated with the system of paths, where

$$H_r^i \underset{\text{def}}{=} \frac{\partial H^i}{\partial \dot{x}^r}.$$

Proof. We must show that a necessary and sufficient condition for

$$H^i(x; \xi) = G^i(x; \xi) \underset{\text{def}}{=} \tfrac{1}{2}\Gamma^i_{jk}\xi^j\xi^k$$

is the invariance of $F(x; \xi)$ under the parallel displacement belonging to H_r^i. The necessity of this condition is evident, since, by Theorem 15.7.1, the parallel displacement associated with $H_r^i = G_r^i$ is length-preserving. To prove sufficiency, assume that F is invariant under the parallel displacement H_r^i, that is, assume that

$$\frac{dF^2}{dt} = \frac{\partial F^2}{\partial x^j} \dot{x}^j + \frac{\partial F^2}{\partial \xi^r} \frac{d\xi^r}{dt} = \left(\frac{\partial F^2}{\partial x^j} - \frac{\partial F^2}{\partial \xi^r} H_j^r\right) \dot{x}^j = 0 \tag{15.19}$$

holds; we shall show that this implies $H^i = G^i$.

We note first that (15.19) holds for all $\dot{x}^j$ and can also be written in the form

$$\frac{\partial g_{ik}}{\partial x^j} \xi^i \xi^k - 2g_{ir} \xi^i H_j^r = 0. \tag{15.20}$$

Let D_r be defined by

$$D_r \underset{\text{def}}{=} \frac{\partial}{\partial \xi^r} \left(\frac{\partial g_{ik}}{\partial x^j} \xi^i \xi^k \xi^j\right);$$

we shall compute D_r in two different ways. On the one hand,

$$D_r = 2 \frac{\partial g_{ir}}{\partial x^j} \xi^i \xi^j + \frac{\partial g_{ij}}{\partial x^r} \xi^i \xi^j. \tag{15.21}$$

(The expression resulting from differentiation of $\partial g_{..}/\partial x_.$ vanishes, by Euler's relation.) On the other hand, by (15.20),

$$D_r = 2 \frac{\partial}{\partial \xi^r} (g_{il} \xi^i H_j^l \xi^j) = 4 \frac{\partial}{\partial \xi^r} (g_{il} \xi^i H^l) = 4g_{rl} H^l + 4g_{il} \xi^i H_r^l,$$

because the homogeneity relation gives $H_j^l \xi^j = 2H^l$. Again, by (15.20), this becomes

$$D_r = 4g_{rl} H^l + 2 \frac{\partial g_{ij}}{\partial x^r} \xi^i \xi^j. \tag{15.22}$$

Equating the right-hand sides of (15.21) and of (15.22), we obtain

$$4g_{rl} H^l = \left(\frac{\partial g_{ir}}{\partial x^j} + \frac{\partial g_{jr}}{\partial x^i}\right) \xi^i \xi^j - \frac{\partial g_{ij}}{\partial x^r} \xi^i \xi^j \tag{15.23}$$

or, after inner multiplication by g^{kr} and division by 2:

$$2H^k = \Gamma^k_{ij}\, \xi^i\, \xi^j.$$

This is what we had to prove.

The theorem can be stated somewhat differently by introducing the holonomy group of the connection and by expressing the invariance in terms of the indicatrix, rather than in terms of the metric base function:

Theorem 15.8.2. *The paths of a system* $\ddot{x}^i + H^i(x; \dot{x}) = 0$ *are the geodesics of a Finsler metric if and only if the holonomy group of the connection* $H^i_r(x; \xi)$ *takes the indicatrix of the metric onto itself.*

Note that this holonomy group consists of homogeneous mappings of the tangent space onto itself that are, in general, nonlinear. The case in which they are linear is of particular interest; it is the case in which the paths are the geodesics of an affine connection. If so, then the question arises, under what conditions does there exist a Riemannian metric whose Christoffel symbols are the coefficients of the affine connection? The answer follows immediately from Theorem 15.8.2:

Theorem 15.8.3. *An affine connection* $\Lambda^i_{jk}(x)$ *is the compatible affine connection belonging to a Riemannian metric* $g_{ik}(x)$, *that is:* $\Lambda^i_{jk} = \Gamma^i_{jk}$ *(the Christoffel symbols), if and only if the holonomy group of* Λ^i_{jk} *at a point* (x_0) *leaves the pointwise Euclidean metric* $g_{ik}(x_0)\xi^i\, \xi^k$ *invariant, that is, takes the indicatrix* $g_{ik}(x_0)\xi^i\, \xi^k = 1$ *onto itself.*

This theorem also answers the question as to the conditions under which a space with an affine connection can be mapped onto a Riemannian space by an affine mapping [in the sense of §13.2(f), that is by a mapping that maps geodesics and their natural parameters onto the like]. We see that this is by no means always possible, but only if the holonomy group is a rotation group when a suitable coordinate system — with $g_{ik}(x_0) = \delta_{ik}$ — is chosen.

Exercise 15.8.4. For $n = 2$, find affine connections whose coefficients are not the Christoffel symbols of any Riemannian metric.

Exercise 15.8.5. *Prove*: If the holonomy group of a system of paths takes every direction of the tangent space onto every other direction, then there exists, to within constant factors, at most one Finsler metric whose geodesics are the given paths. [This theorem, as well as the others of this subsection, are to be found in *Ann. Univ. Saraviensis, Scientia* **5**, 235–244 (1956).]

15.9 Schematic Survey of the Types of Spaces and of Their Characterizations

We close this chapter by giving a summary view of the space structures considered in this book. We briefly characterize the basic structure of each of the spaces and indicate geometric properties that distinguish it among the more general types:

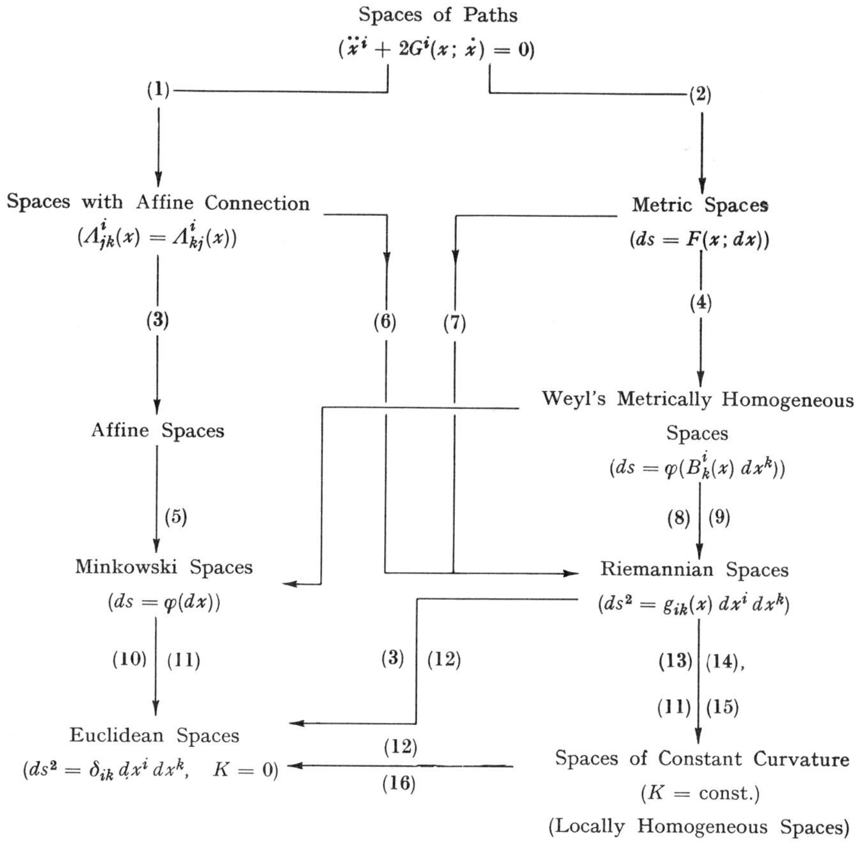

The numbers in this schema refer to the following characteristic properties:

(1) Infinitesimally affine spaces: At every point there exists a coordinate system such that $G^i = 0$ at the point.

(2) The holonomy group of G^i_r leaves an indicatrix invariant (Theorem 15.8.2).

(3) Parallel displacement is independent of path (§10.3).

(4) Metric homogeneity: The indicatrices at different points all are images of a fixed one of them under linear transformations (§15.4).

(5) The metric is invariant under the parallel displacements of the affine space: Parallel vectors have the same length.

(6) The holonomy group of the affine connection, in a suitable coordinate system, consists of orthogonal transformations (Theorem 15.8.3).

(7) Pointwise isotropy (Theorem 15.5.1).

(8) Pointwise free mobility (§15.3).

(9) Theorem of Weyl on the existence of affine connections (§15.4).

(10) Isotropy (Theorem 15.5.1).

(11) Free mobility of rigid bodies of finite extension (§15.3).

(12) Existence of a similarity mapping (Theorem 13.4.2).

(13) Infinitesimal isotropy (Schur's Theorem, 12.2.2).

(14) Homogeneity ($K =$ const.).

(15) Existence of a geodesic mapping into Euclidean space (Beltrami's Theorem, 13.5.1).

(16) Euclidean axiom of parallels (in its local version: In geodesic triangles, the sum of the angles equals 2π).

We emphasize that the above schematic survey cannot make any claims of completeness, either as to the types of spaces that are included or as regards their characterizing properties. We had to restrict ourselves to the results discussed in this book. For further differential-geometric types of spaces, we refer the reader particularly to J. A. Schouten, "Ricci Calculus," 2nd ed., Berlin-Göttingen-Heidelberg, 1954 (Springer "Grundlagen" series, Vol. 10), which is a comprehensive encyclopedia of the tensor calculus and its applications.

CHAPTER V

Selections from Differential Geometry in the Large

In this chapter we shall study closed curves and surfaces in their total extension, that is, "globally" (or, "in the large"). In contradistinction to differential geometry in the small, there are for global geometry no uniform recipe-like methods of procedure available as yet. Rather, different problems mostly require entirely individual methods of proof. This gives the global problems a special attraction which is heightened further by the beauty of the results. There is one method, however, which was first used systematically by G. Herglotz and which has since been developed into a powerful tool; this is the so-called *method of integral formulas*. It can be described briefly as follows: One shows first that a sufficient condition for some global property is the identical vanishing of a certain function F; one proves next that F cannot change sign anywhere in the manifold, and, finally, that the integral of F (or of $F \cdot G$, where G is a function of fixed sign) vanishes; this establishes that $F \equiv 0$ and thus proves the desired result. We shall use mainly this method of Herglotz, but, on occasion, other methods as well.

§ 16. Curves in the Large

16.1 Plane Convex Curves

A space curve $\mathfrak{x}(t)$, $-\infty < t < +\infty$, is called *closed* if there is a real number p such that $\mathfrak{x}(t + p) = \mathfrak{x}(t)$. A closed plane curve is called *convex* if every straight line has at most two points in common with the curve. A convex curve lies entirely on one side of any one of its tangents, for, if a tangent had a second point in common with the curve, then a neighboring straight line would have more than two points in common with the curve, against the assumption of convexity. A plane convex curve is called an *oval* if it is of class C^2 with respect to arc length and satisfies $\kappa(s) = |\mathfrak{x}''(s)| > 0$ for all s. An oval always curves toward one and the same side; its principal normal vector $\mathfrak{v}_2$, which has the same direction as $\mathfrak{x}''$, always points toward the inside region. It is expedient to assume the coordinate origin to be inside the oval. Let $\alpha(s)$ be the angle from a fixed direction, say the positive x_1-direction, to the principal normal vector. Then

$$\left|\frac{d\alpha}{ds}\right| = \kappa(s) > 0 \tag{16.1}$$

insures that the transformation $\alpha = \alpha(s)$ is an admissible parameter transformation (see Fig. 42). If $\mathfrak{y}$ stands for the general point of the tangent line to the oval at $\mathfrak{x}(\alpha)$, the equation of this tangent line is

$$(\mathfrak{y} - \mathfrak{x}(\alpha))\mathfrak{v}_2(\alpha) = 0 \tag{16.2}$$

or

$$y_1 \cos\alpha + y_2 \sin\alpha + h(\alpha) = 0, \tag{16.3}$$

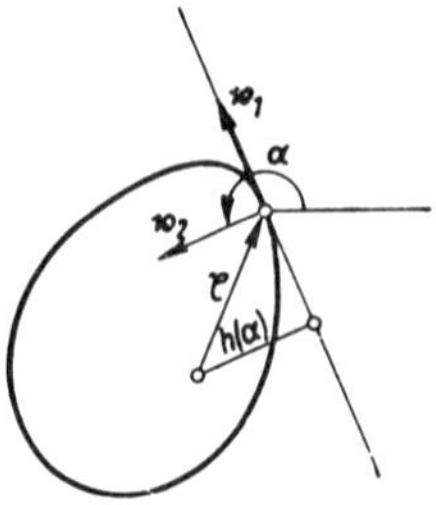

FIG. 42. Oval.

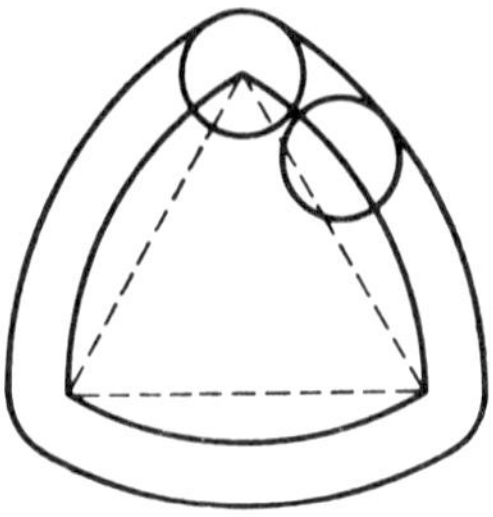

FIG. 43. Reuleaux triangle and parallel curve.

where

$$h(\alpha) = -\,\mathfrak{x}(\alpha)\mathfrak{v}_2(\alpha) \tag{16.4}$$

is the distance of the tangent line from the origin. The function $h(\alpha)$ is Minkowski's *support function* of the oval. While this function lacks intrinsic geometric meaning because it depends on the position of the origin, it can be used to express geometric quantities such as, for instance, the distance between two parallel tangents,

$$h(\alpha) + h(\alpha + \pi) = b(\alpha), \tag{16.5}$$

which is called the *width* of the curve in the corresponding direction (i.e., the direction perpendicular to the two tangents). In general, this width depends on α. However, the circle is not the only curve of constant width, $b(\alpha) = b = \text{const.}$ (see Exercise 16.1.2). The curves of constant width all have the following important property in common with the circle:

Theorem 16.1.1. (Barbier, 1860). *All curves of constant width b have the same circumference πb.*

Proof. We shall first express the curvature in terms of the support function. We denote differentiations with respect to α by dots, with respect to s by primes. From $\mathfrak{v}_2 = (\cos\alpha, \sin\alpha)$, we obtain $\dot{\mathfrak{v}}_2 = (-\sin\alpha, \cos\alpha)$ and $\ddot{\mathfrak{v}}_2 = -\mathfrak{v}_2$. Therefore, differentiation of $\mathfrak{x}\mathfrak{v}_2 = -h$ yields $\mathfrak{x}\dot{\mathfrak{v}}_2 = -\dot{h}$, whence $-\mathfrak{x} = h\mathfrak{v}_2 + \dot{h}\dot{\mathfrak{v}}_2$, and

$$-\mathfrak{x}'\frac{1}{\kappa} = -\mathfrak{x}'\frac{ds}{d\alpha} = -\dot{\mathfrak{x}} = \dot{h}(\mathfrak{v}_2 + \ddot{\mathfrak{v}}_2) + (h + \ddot{h})\,\dot{\mathfrak{v}}_2 = (h + \ddot{h})\,\dot{\mathfrak{v}}_2$$

or

$$\kappa = \frac{1}{|h + \ddot{h}|}. \tag{16.6}$$

Now the circumference of the curve is obtained from (16.6) and (16.5), with $b(\alpha) = b$, as

$$\int_{\alpha=0}^{\alpha=2\pi} ds = \int_0^{2\pi} \frac{ds}{d\alpha}\,d\alpha = \int_0^{2\pi} (h + \ddot{h})\,d\alpha = \int_0^{2\pi} h\,d\alpha = \int_0^{\pi} (h(\alpha) + h(\alpha + \pi))\,d\alpha = \pi b,$$

since the integral of a derivative $\ddot{h} = (\dot{h})^{\cdot}$ vanishes.

This theorem is a global theorem because the hypothesis, as well as the conclusion, contains statements dealing with the curve in its total extent.

Exercise 16.1.2. *Prove* that the following are curves of constant width:

(a) The *Reuleaux triangle,* obtained from an equilateral triangle by replacing each side with an arc of a circle centered at the opposite vertex.

(b) The (smooth) parallel curves of the Reuleaux triangle, obtained as the envelopes of all circles of fixed radius r whose centers range over the Reuleaux triangle (see Fig. 43).

16.2 The Four-Vertex Theorem

This is probably the best known of all theorems about curves in the large. It was proved in 1909 by S. Mukhopadhyaya and independently in 1912 by Adolf Kneser. The theorem is concerned with the minimum number of *vertices*, that is, of points where κ' changes sign,[1] on an oval whose curvature

[1] In the weak sense, that is, from nonpositive to nonnegative or vice versa.

is differentiable. On the ellipse there are exactly four vertices; on the circle, every point is a vertex.

Four-Vertex Theorem. *On every oval with differentiable curvature, there are at least four vertices.*

Proof. The continuous function $\kappa(s)$ must assume its maximum and its minimum values on the closed oval and thus we already have two vertices. Through these two, we pass a straight line. In a coordinate system with its origin on the chord joining the two vertices, the straight line has an equation $\mathfrak{c}\mathfrak{x} = 0$. We shall first establish an integral formula by using integration by parts, noting in the process that the terms evaluated at the limits cancel out because the curve is closed, and making use of the second Frenet formula (all integrals to be taken around the entire oval):

$$\oint \kappa' \mathfrak{x} \, ds = -\oint \kappa \mathfrak{x}' \, ds = -\oint \kappa \mathfrak{v}_1 \, ds = \oint \mathfrak{v}_2' \, ds = 0 \qquad (16.7a)$$

and, therefore, also

$$\oint \kappa' \mathfrak{c}\mathfrak{x} \, ds = 0. \qquad (16.7b)$$

If there were no further vertex, then, because $\oint \kappa' ds = 0$, the quantity κ' would have to be positive on one and negative on the other of the two arcs of the oval, giving $\kappa' \mathfrak{c}\mathfrak{x}$ the same sign on both arcs and thus leading to a contradiction against (16.7b). Therefore, there is at least a third vertex and, in fact, there must then be a fourth as well, since, otherwise, the chord joining certain two of the three vertices would again subdivide the oval into two arcs on each of which κ' retains the same sign.

This proof, due to Herglotz, is an example of the method of integral formulas.

However, this short proof of Herglotz yields somewhat less than Kneser's proof. Kneser proved the four-vertex theorem for arbitrary — not necessarily convex — plane curves free of double points (i.e., of self-intersections). Kneser and Mohrmann (1917) also gave a type of four-vertex theorem for certain kinds of space curves, as follows: On a closed space curve free of double points and lying on the bounding surface of a convex body, there are at least four points where the osculating plane is stationary. [The first complete proof is to be found in M. Barner and F. Flohr, Der Vierscheitelsatz und seine Verallgemeinerungen, Der Mathematikunterricht 4, No. 4 (1958); they also present many other theorems about curves in the large, as well as a voluminous bibliography for these topics. — There do exist, incidentally, closed space curves free of double points for which $\kappa' = 0$ at two points only; see H. Gericke, *Jahresber. DMV* **47**, 22–24 (1937).]

16.3 Antipodal Pairs of Points on Ovals

Two points of an oval are called an *antipodal pair* if the tangents at the two points are parallel and the curvatures equal. Hence, by the formulas in § 16.1, and if we set $1/\kappa(\alpha) = \rho(\alpha)$, the point $\mathfrak{x}(\alpha)$ belongs to an antipodal pair if, and only if,

$$F(\alpha) \underset{\text{def}}{=} \rho(\alpha) - \rho(\alpha + \pi) = 0. \tag{16.8}$$

The following result is due to W. Blaschke and W. Süss:

Theorem 16.3.1. *On every oval there are at least three antipodal pairs of points.*

Proof. We must prove that $F(\alpha)$ has at least three zeros in $0 \leqq \alpha < \pi$. If $\mathfrak{e}_1$ is the unit vector in the x_1-direction, then

$$\int_0^{2\pi} \rho(\alpha) \sin \alpha \, d\alpha = \int_0^{2\pi} \frac{ds}{d\alpha} \mathfrak{e}_1 \mathfrak{v}_1 \, d\alpha = \mathfrak{e}_1 \int_0^{2\pi} \frac{d\mathfrak{x}}{ds} \, ds = 0,$$

where we made use of $\sin \alpha = \mathfrak{e}_1 \mathfrak{v}_1$. Now, since $F(\alpha + \pi) = - F(\alpha)$, the function $F(\alpha)$ has at least two zeros (even sign changes), α_0 and $\alpha_0 + \pi$. We choose coordinates such that $\alpha_0 = 0$. Now assume, say, $F(\alpha) \geqq 0$ (the argument would be similar for $\leqq 0$) in some right-hand neighborhood of $\alpha = 0$; then $F(\alpha) \geqq 0$ also in a left-hand neighborhood of $\alpha = \pi$. By (16.8),

$$\int_0^{\pi} F(\alpha) \sin \alpha \, d\alpha = \int_0^{\pi} \rho(\alpha) \sin \alpha \, d\alpha - \int_0^{\pi} \rho(\alpha + \pi) \sin \alpha \, d\alpha = \int_0^{2\pi} \rho(\alpha) \sin \alpha \, d\alpha = 0,$$

so that F cannot be nonnegative in the entire interval $0 \leqq \alpha \leqq \pi$, unless it vanishes identically; thus F must have, in any case, two further zeros in the interior of that interval and the theorem is proved.

Exercise 16.3.2. On the circle and on the ellipse — in fact, on any curve with a center — every point is a member of an antipodal pair. *Find* an oval with exactly three antipodal pairs of points. (*Hint.* Find an oval with the symmetries of the equilateral triangle and such that the curvature is monotonic on each of the six congruent subarcs.)

16.4 The Total Curvature of Closed Space Curves

We consider next an entirely different type of problem, a problem first tackled by W. Fenchel (1929) and completely settled by him, namely: For a closed space curve, the integral

$$\oint \kappa(s)\, ds$$

is always positive; can we give a sharp lower bound for it? Since $\kappa(s)\, ds = |\mathfrak{x}''|\, ds = |d\mathfrak{v}_1|$, the value of the integral is simply the total length of the spherical tangent image $\mathfrak{v}_1(s)$ of the given curve. Thus the following result seems all the more remarkable:

Fenchel's Theorem. *For every closed space curve,*

$$\oint \kappa(s)\, ds \geqq 2\pi. \tag{16.9}$$

Proof. Many proofs have been given for this theorem since 1929. The one we shall present is by K. Voss [*Arch. Math.* **6** (1955)] and is particularly transparent. We let the given closed curve $\mathfrak{x}(s)$ be the center line of a tubular surface (see §6.7); this surface we can construct by letting a circle in the normal plane, with center on the curve $\mathfrak{x}(s)$ and of sufficiently small radius ε, travel so that its center ranges over the whole curve. We shall prove first that, for this tubular surface (as, indeed, for every closed surface), the following holds true: the integral $\int K\, dF$, taken over that part of the surface where $K \geqq 0$, has a value of 4π at least:

$$\int_{K \geqq 0} K\, dF \geqq 4\pi. \tag{16.10}$$

To prove this, note that $K\, dF$ is the element of surface area of the spherical normal image (§5.3), and we need only prove that the spherical normal image of the relevant portions (with $K \geqq 0$) of the closed surface covers the unit sphere — whose area is 4π — completely; that is, we need only show that every unit vector $\mathfrak{N}_0$ appears as the unit normal to the given surface at least once, at a point where $K \geqq 0$. Now, if a plane perpendicular to $\mathfrak{N}_0$ is moved from infinitely far out up to its first position of contact with the surface (which is closed and, therefore, also bounded), then the normal vector at any one of the points of intersection is $\mathfrak{N}_0$ and, as the surface lies entirely on one side of the tangent plane, we also have $K \geqq 0$.

Next, (16.9) follows from (16.10) by direct computation. Let us parametrize the surface by taking for u^1 the arc length s of its center line (the given curve), and for u^2 the angle made by the relevant radius of the circle with the principal normal of the curve. Then the tubular surface is given by

$$\mathfrak{x}(u^1, u^2) = \mathfrak{x}(u^1) + \varepsilon(\mathfrak{v}_2(u^1) \cos u^2 + \mathfrak{v}_3(u^1) \sin u^2).$$

In calculating the fundamental quantities of the surface, we make use of the Frenet formulas to substitute for all derivatives of the $\mathfrak{v}_i$ that occur, and we obtain

$$\mathfrak{x}_1 \times \mathfrak{x}_2 = \varepsilon(\kappa\varepsilon \cos u^2 - 1)(\mathfrak{v}_2 \cos u^2 + \mathfrak{v}_3 \sin u^2),$$

so that, if $\varepsilon < \min 1/\kappa$, we actually do have a surface; also,

$$\mathfrak{N} = - \mathfrak{v}_2 \cos u^2 - \mathfrak{v}_3 \sin u^2.$$

From this, we get $K\, dF = (L/g) \sqrt{g}\, du^1\, du^2 = - \kappa \cos u^2\, du^1\, du^2$, and if we exclude $\kappa = 0$ for the time being (leaving it to the reader to carry the argument through for such points), we have $K \geqq 0$ if $\cos u^2 \leqq 0$, hence

$$4\pi \leqq \int_{K \geqq 0} K\, dF = - \iint_{\cos u^2 \leqq 0} \kappa(u^1) \cos u^2\, du^1\, du^2 = 2 \oint \kappa\, du^1,$$

which is what we wanted to prove.

The total torsion of a space curve can assume any positive or negative value at all. For spherical curves, however, one can prove that $\oint \tau\, ds = 0$. W. Scherrer proved in 1940 that spherical and plane surfaces are the only ones on which all curves have vanishing total torsion.

Exercise 16.4.1. *Prove* that $\oint \kappa\, ds = 2\pi$ for all plane convex curves, and for no others.

Exercise 16.4.2. Theorem of Fàry (1949) and Milnor (1950): *Prove* that, if a closed space curve is knotted, then $\oint \kappa\, ds \geqq 4\pi$.

[For both problems, see the paper by Voss cited in this subsection, also W. Fenchel, *Bull. Amer. Math. Soc.* **57**, 44–45 (1951).]

§ 17. Surfaces in the Large

17.1 Characteristic Properties of the Sphere

A closed surface in Euclidean 3-space is complete and compact, hence, in particular, bounded. An important class of closed surfaces is constituted by all those for which the Gaussian curvature is everywhere nonnegative;

this is the class of *ovaloids*. An ovaloid always lies entirely on one side of its tangent plane and is convex, meaning that it meets every straight line at most in a closed segment (which may reduce to a point, or be empty). Examples of closed surfaces are the torus and, more generally, any tubular surface belonging to a closed space curve as its center line. Examples of ovaloids are spheres and ellipsoids.

The spheres are characterized by various special properties among all closed surfaces. We begin with a lemma of local character.

Theorem 17.1.1. *On a surface with $K > 0$ and not (part of) a sphere, let Q be a point at which the larger principal curvature κ_1 assumes a maximum. Then the smaller principal curvature κ_2 cannot assume a minimum at Q.*

Proof. We give an indirect proof. Assume there were a point 0 such that

$$\kappa_1(0) \geqq \kappa_1(P), \qquad \kappa_2(0) \leqq \kappa_2(P), \qquad (\kappa_1(P) \geqq \kappa_2(P))$$

for all points P of the surface. If 0 is an umbilic, with $\kappa_1(0) = \kappa_2(0)$, then it follows that all other points P are likewise umbilics and the surface is part of a sphere (Theorem 6.2.1). In the contrary case there is a neighborhood of 0 in which the lines of curvature can be made the parametric curves in such a way that κ_j is the principal curvature along the u^j-curves and that along the parametric curves through 0, the parameter is the arc length. In this neighborhood we then have $g_{12} = 0 = L_{12}$, and we also have $g_{11}(0) = g_{22}(0) = 1$. For the principal curvatures we have

$$\kappa_j = L_{jj}/g_{jj},$$

and the equations of Gauss and of Mainardi and Codazzi in line-of-curvature parameters (see Exercises 5.5.4 and 5.5.6) yield

$$K = \frac{L_{11}\,L_{22}}{g_{11}\,g_{22}} = -\frac{1}{2\sqrt{g}}\left[\left(\frac{g_{11,2}}{\sqrt{g}}\right)_{,2} + \left(\frac{g_{22,1}}{\sqrt{g}}\right)_{,1}\right]$$

and

$$L_{11,2} = \tfrac{1}{2}\, g_{11,2}(\kappa_1 + \kappa_2), \qquad L_{22,1} = \tfrac{1}{2}\, g_{22,1}(\kappa_1 + \kappa_2);$$

therefore

$$\kappa_{1,2} = \tfrac{1}{2}\,\frac{g_{11,2}}{g_{11}}\,(-\,\kappa_1 + \kappa_2), \qquad \kappa_{2,1} = \tfrac{1}{2}\,\frac{g_{22,1}}{g_{22}}\,(\kappa_1 - \kappa_2) \tag{17.1}$$

and

$$g_{11,2} = -\,\frac{2g_{11}\,\kappa_{1,2}}{\kappa_1 - \kappa_2}, \qquad g_{22,1} = \frac{2g_{22}\,\kappa_{2,1}}{\kappa_1 - \kappa_2}. \tag{17.2}$$

If we substitute (17.2) into the expression for K, it follows that

$$K = -\frac{1}{\sqrt{g}}\left[\left(\frac{-g_{11}\,\kappa_{1,2}}{\sqrt{g}(\kappa_1-\kappa_2)}\right)_{,2} + \left(\frac{g_{22}\,\kappa_{2,1}}{\sqrt{g}(\kappa_1-\kappa_2)}\right)_{,1}\right] \tag{17.3}$$

$$= \frac{1}{\sqrt{g}}\,\frac{g_{11}\,\kappa_{1,22} - g_{22}\,\kappa_{2,11}}{\kappa_1-\kappa_2} + a\cdot\kappa_{1,2} + b\cdot\kappa_{2,1},$$

where the functions a and b are of no interest here. By the assumed extremal properties of the principal curvatures at 0, we have at this point that

$$\kappa_{1,2} = \kappa_{2,1} = 0, \qquad \kappa_{1,22} \leqq 0, \qquad \kappa_{2,11} \geqq 0,$$

which finally yields $K(0) \leqq 0$ and thus a contradiction. This completes the proof. The theorem itself goes back to Hilbert; a sharper version was proved by S. S. Chern in 1945 [see W. Blaschke, "Differentialgeometrie I," p. 197, and K. Voss, *Math. Ann.* **131**, 181 (1956)].

The local Theorem 17.1.1 permits a number of global applications.

Theorem 17.1.2. *The spheres are the only closed surfaces of constant Gaussian curvature,* $K = K_0$.

Proof. First, we must have $K_0 \geqq 0$, since on every closed surface there are points with $K \geqq 0$, for instance, the points with the largest x_1-coordinate. $K = 0$ is not possible, since the complete surfaces with $K = 0$ all are ruled surfaces which cannot be bounded, hence, certainly not closed. If $K = K_0 > 0$, then at any point where κ_1 is a maximum (such points exist because of compactness), κ_2 is a minimum because $\kappa_2 = K_0/\kappa_1$. Now we are through, by Theorem 17.1.1.

The *theorema egregium* shows that a closed surface isometric to the sphere must itself be of constant Gaussian curvature. This gives the following corollary of Theorem 17.1.2:

Theorem 17.1.3. *The (complete) sphere cannot be bent.*

This result[1] was conjectured by Minding in 1838, partly proved by Jellet in 1854, and proved completely by H. Liebmann in 1899.

[1] In this connection it is of interest that a sphere with a hole cut into it, no matter how small the hole, *can* be bent (tennis ball). This was proved, even for arbitrary ovaloids — which cannot be bent, as we shall prove in §17.3 — independently by W. Süss and S. Cohn-Vossen in 1927, for infinitesimal bendings, and by Pogorelov (1951) and G. Hellwig for finite bendings.

The next result goes substantially further, although it restricts the field to ovaloids with $K > 0$ from the beginning. It is due to A. D. Aleksandrov (1938) and S. S. Chern (1954):

Theorem 17.1.4. *If a relation $\kappa_2 = f(\kappa_1)$ holds on an ovaloid where f is a monotonically decreasing function, then the ovaloid is a sphere.*

Proof. By the monotonicity of f, any point which gives a maximum for κ_1 also gives a minimum for κ_2. Since $\kappa_1 \kappa_2 > 0$, the assertion then follows from Theorem 17.1.1.

In particular, since $\kappa_2 = 2H - \kappa_1$ we have the following corollary:

Theorem 17.1.5. *The spheres are the only ovaloids whose mean curvature H is constant* (H. Liebmann, 1900).

Theorem 17.1.2 holds for all closed surfaces and thus there do not exist, for instance, surfaces of the topological type of the torus on which $K =$ const. On the other hand, the only closed surfaces considered at all for Theorems 17.1.4 and 17.1.5 are ovaloids with $K > 0$ to begin with. However, A. D. Aleksandrov has shown recently that the spheres are also the only surfaces with $H =$ const. among all closed surfaces free of self-intersections. For partial results, in particular by H. Hopf, see the paper by K. Voss cited in this subsection. If gravity is neglected, surfaces with $H =$ const. can be physically realized as soap films with constant air pressure on both sides; hence the only possible closed soap films (soap bubbles) are spheres.

17.2 Further Development of the Tensor Calculus on Surfaces

In the theory of surfaces, we occasionally encountered, besides the fundamental tensors g_{ik} and L_{ik}, the tensor ε_{ik}, defined as follows:

$$\varepsilon_{11} = \varepsilon_{22} = 0, \qquad \varepsilon_{12} = -\varepsilon_{21} = \sqrt{g}. \tag{17.4}$$

Before continuing with our study of surfaces in the large, we shall go into some auxiliary considerations involving this ε-tensor.

By inner multiplication, we obtain

$$\varepsilon^{i}_{.k} \underset{\text{def}}{=} g^{ij}\,\varepsilon_{jk}; \qquad \varepsilon^{ik} \underset{\text{def}}{=} g^{ij}\,g^{kl}\,\varepsilon_{jl}; \qquad \varepsilon_{k}^{\cdot i} = g^{ij}\,\varepsilon_{kj} \tag{17.5}$$

and by a calculation, we find

$$\varepsilon^{11} = \varepsilon^{22} = 0; \qquad \varepsilon^{12} = -\varepsilon^{21} = \frac{1}{\sqrt{g}} \tag{17.6}$$

and

$$\varepsilon^{ik}\,\varepsilon_{jk} = \delta^{i}_{j}. \tag{17.7}$$

Since the components of the ε-tensors depend only on the g_{ik}, it follows from Ricci's lemma that

$$\varepsilon_{ik;j} = 0; \qquad \varepsilon_{k;j}^{.i} = \varepsilon_{.k;j}^{i} = 0; \qquad \varepsilon^{ik}{}_{;j} = 0. \tag{17.8}$$

Like every (1, 1)-tensor, $\varepsilon^{i}_{.k}$ can be interpreted as a mapping of the tangent plane onto itself; specifically, the vector $*\xi^i \underset{\text{def}}{=} \varepsilon_k^{.i}\, \xi^k$ is obtained from the vector ξ^k by a rotation through 90°. This can easily be verified, using special parameters in which $g_{11} = g_{22} = 1$, $g_{12} = 0$, at the point in question. We can use the ε-tensors to rewrite formula (4.13) for the geodesic curvature κ_g, as follows:

$$\kappa_g = \varepsilon_{ki}\, u^{k'} \frac{Du^{i'}}{ds} = g_{ij}\, \varepsilon_k^{.j}\, u^{k'} \frac{Du^{i'}}{ds} = g_{ij}\, h^j \frac{Du^{i'}}{ds}, \tag{17.9}$$

where

$$h^j \underset{\text{def}}{=} \varepsilon_k^{.j}\, u^{k'}$$

is that unit vector in the surface that is obtained from the tangent vector by a rotation through 90°. (Note that the sign depends on the direction of $\mathfrak{N}$.)

We can also use the ε-tensor to assign to every doubly covariant tensor a_{jl} an "adjoint" tensor, as follows:

$$*a^{ik} \underset{\text{def}}{=} \varepsilon^{ij}\, \varepsilon^{kl}\, a_{jl}. \tag{17.10}$$

A computation, using the definitions, yields

$$*a^{ij}\, a_{kj} = \frac{a}{g}\, \delta_k^i, \qquad a \underset{\text{def}}{=} \det(a_{ik}). \tag{17.11}$$

We can now contract over i and k and obtain a convenient formula for the determinant a of a tensor a_{ik}:

$$2\frac{a}{g} = *a^{ij}\, a_{ij}. \tag{17.12}$$

In particular, we obtain for the Gaussian curvature K that

$$K = \frac{L}{g} = \frac{1}{2}\, *L^{ij}\, L_{ij}. \tag{17.13}$$

The ε-tensor also proves to be very useful in providing short notations for expressions of the type $T_{ik..} - T_{ki..}$ such as occur, for instance, in integrability conditions. Thus, for example, we can write the Mainardi-Codazzi equations (11.22) in the form

$$\varepsilon^{ik} L_{jk;i} = 0 \qquad \text{or} \qquad {}^*L^{ji}{}_{;i} = 0, \tag{17.14}$$

while the *theorema egregium* can be written

$$K = -\tfrac{1}{4}\,\varepsilon^{ij}\,\varepsilon^{kl}\,R_{ijkl}. \tag{17.15}$$

Theorem 9.6.1, which deals with changing the order of covariant differentiations, assumes the form

$$\varepsilon^{kj}\,v_{i;j;k} = K\varepsilon_i^{\cdot r}\,v_r. \tag{17.16}$$

Expressions involving differences occur also in the integral theorems of Gauss and Stokes. Here too we can make use of the ε-tensor. Let $\mathfrak{G}$ be a (simple) region on a surface, with a smooth boundary $\partial\mathfrak{G}$ which is to be traversed in such a way that the region, when viewed from the positive direction of the surface normal, always lies on the left of the boundary. Let $v_k(u^i)$ be a vector field continuous on $\mathfrak{G}$ plus its boundary and differentiable within $\mathfrak{G}$. To it — using first the corresponding region of the (u^1, u^2)-plane — we can apply Gauss' Theorem (also called Green's Theorem in the plane), obtaining

$$\iint\limits_{\mathfrak{G}} (v_{2,1} - v_{1,2})\,du^1\,du^2 = \oint\limits_{\partial\mathfrak{G}} v_i \frac{du^i}{ds}\,ds.$$

Since the Γ are symmetric, we can replace ordinary by covariant derivatives and if we also introduce $dF = \sqrt{g}\,du^1\,du^2$, we obtain an invariant version of the theorem,

$$\iint\limits_{\mathfrak{G}} \varepsilon^{ik}\,v_{k;i}\,dF = \oint\limits_{\partial\mathfrak{G}} v_k\,u^{k'}\,ds. \tag{17.17}$$

Analogously, with $h_i = \varepsilon_{ki}\,u^{k'}$ we can prove the integral formula

$$\iint\limits_{\mathfrak{G}} v^k{}_{;k}\,dF = -\oint\limits_{\partial\mathfrak{G}} v^k\,h_k\,ds. \tag{17.18}$$

The integral formulas (17.17) and (17.18) are the basis on which rests the derivation of further integral formulas in the theory of surfaces.

17.3 The Integral Formula of Gauss and Bonnet

Before deriving this formula, the oldest integral formula of the theory of surfaces, we shall discuss a more general formula. Let $\mathfrak{G}$ be a region on a surface of class C^3, with boundary curve $\partial\mathfrak{G}$ of class C^2. Let v_i be a vector field of class C^2 on $\mathfrak{G} \cup \partial\mathfrak{G}$. It follows from (17.12) and (17.8) that

$$2\frac{\det(v_{i;k})}{g} = \varepsilon^{ij}\,\varepsilon^{kl}\,v_{i;k}\,v_{j;l} = \varepsilon^{kl}(\varepsilon^{ij}\,v_{j;l}\,v_i)_{;k} - \varepsilon^{kl}\,\varepsilon^{ij}\,v_{j;l;k}\,v_i.$$

From (17.16), we have $\varepsilon^{kl}\,v_{j;l;k} = K\varepsilon_j^{\cdot r}\,v_r$, and therefore

$$2\frac{\det(v_{i;k})}{g} = \varepsilon^{kl}(\varepsilon^{ij}\,v_{j;l}\,v_i)_{;k} - K\varepsilon^{ij}\,\varepsilon_j^{\cdot r}\,v_i\,v_r,$$

and since $\varepsilon^{ij}\,\varepsilon_j^{\cdot r} = -\,g^{ir}$, we obtain from Gauss' theorem (17.17), using $g^{ir}\,v_i\,v_r = v^2$, that

$$2\iint\limits_{\mathfrak{G}} \frac{\det(v_{i;k})}{g}\,dF = \oint\limits_{\partial\mathfrak{G}} \varepsilon^{ij}\,v_{j;l}\,v_i\,u^{l'}\,ds + \iint\limits_{\mathfrak{G}} Kv^2\,dF.$$

In the integral over the boundary, we can transform the integrand. Along $\partial\mathfrak{G}$, we have

$$v_i = v(u^{i'}\cos\beta + h^i\sin\beta),$$

where β is the angle between v_i and the tangent to $\partial\mathfrak{G}$. Using the Frenet formulas for curves in a two-dimensional Riemannian space,

$$\frac{Du^{i'}}{ds} = \kappa_g\,h^i, \qquad \frac{Dh^i}{ds} = -\,\kappa_g\,u^{i'}$$

(see §11.4), we obtain

$$\varepsilon^{ij}\,v_{j;l}\,v_i\,u^{l'} = \varepsilon^{ij}\,v_i\,\frac{Dv_j}{ds} = v^2(\kappa_g + \beta'),$$

so that the last integral formula becomes[1]

$$2\iint\limits_{\mathfrak{G}} \frac{\det(v_{i;k})}{g}\,dF = \oint\limits_{\partial\mathfrak{G}} v^2(\kappa_g + \beta')\,ds + \iint\limits_{\mathfrak{G}} v^2\,K\,dF. \tag{17.19}$$

We now prove the following result:

---

[1] For this derivation of (17.9), and for the derivation of the Gauss-Bonnet theorem that follows, I am obliged to K. P. Grotemeyer.

Theorem 17.3.1 (of Gauss and Bonnet). *Let* $\mathfrak{G}$ *be a simply connected*[2] *region of a surface. Then*

$$\iint_{\mathfrak{G}} K\,dF + \oint_{\partial\mathfrak{G}} \kappa_g\,ds = 2\pi. \tag{17.20}$$

(*The surface integral is called the* total curvature — *or, following Gauss, the* curvatura integra — *of the surface region.*)

Proof. We choose on $\mathfrak{G}$ (including its boundary) a field of unit vectors, so that $v^2 = 1$. (On a simply connected[2] region, such a vector field always exists: In the special coordinates, set $w^1 = 1$, $w^2 = 0$, say, and

$$v^i = w^i / \sqrt{g_{kj}\, w^k\, w^j}.\ )$$

Since $v_i v^i = 1$, covariant differentiation yields $v_{i;k}\, v^i = 0$, whence $\det(v_{i;k}) = 0$. Substituting all this into (17.19), we obtain

$$\iint_{\mathfrak{G}} K\,dF + \oint_{\partial\mathfrak{G}} \kappa_g\,ds = -\oint_{\partial\mathfrak{G}} \beta'\,ds. \tag{17.21}$$

Hence it only remains to show that the (signed) contour integral on the right equals 2π. This right-hand side represents the total angle of rotation of the tangent to the curve against the 1-direction and is therefore certainly equal to an integral multiple $n \cdot 2\pi$, since both directions coincide with their initial positions after a full traversal. If the curve is shrunk to a point, this quantity must change continuously; being capable of assuming only integral multiples of 2π as its values, it must, therefore, remain constant. For small circles about a point of the region, it is indeed intuitively clear that $n = 1$, and we shall establish this fact by following a method of proof due to L. Bieberbach and E. Schmidt. If g_{ik} defines a positive-definite metric on our two-dimensional manifold, then so does $(1 - t)\delta_{ik} + tg_{ik}$ (*Exercise*: Prove this) and, if t varies continuously, n remains fixed. We can therefore calculate the value of n relative to the Euclidean metric δ_{ik}. We replace the tangent

---

[2] A region on a surface is said to be *simply connected* if its boundary curve can be shrunk down continuously to any point of the region without, in the process, passing over points that do not belong to the region. Thus the region cannot, in particular, have any holes or singular points. We assume further — this does not restrict generality — that the region of the surface can be mapped one-to-one onto a single region of the plane of parameters, hence, that it can be covered by a single coordinate neighborhood.

line by a chord joining two sufficiently close points of the curve; then we let the initial point of the chord range over a neighboring curve inside the original curve; next we shrink this inner curve to a point; and, finally, we deform the outer curve $\partial\mathfrak{G}$ — the locus of the end point of our chord — into a circle with center at the point to which the inner curve was shrunk. Under all these continuous deformations n remains fixed and the number of rotations of the radius vector of the final circle is indeed 1, so that $n = 1$.

The theorem as stated was given by O. Bonnet in 1848; C. F. Gauss had dealt with the following important limiting case in 1827:

Corollary 1. *Let Δ be a geodesic triangle on a surface, that is a triangle whose sides are geodesics. If α, β, γ are the interior angles of this triangle, then*

$$\iint_{\Delta} K\,dF = (\alpha + \beta + \gamma) - \pi.$$

This formula yields the "spherical excess" and "defect," respectively (with, say, $K = +1$ and $K = -1$), for spherical and for hyperbolic geometry.

Proof. We approximate the triangle from inside, say, by rounding off its corners, by a curve of class C^2 to which we can apply Theorem 17.3.1. By an earlier result (see Exercise 4.7.5), $\kappa_g = d\varphi/ds$, where φ is the angle between the tangent and a vector of a field of geodesic parallels along the curve. Hence $\int \kappa_g\,ds = 0$ along the geodesic arcs, while, along the three rounded-off corners, the integral of κ_g differs arbitrarily little from the exterior angles $\pi - \alpha, \pi - \beta, \pi - \gamma$, respectively, and Corollary 1 is proved.

Corollary 2. *If $\partial\mathfrak{G}$ is piecewise of class C^2 and if the exterior angles at the corners are δ_i, then*

$$\iint_{\mathfrak{G}} K\,dF + \oint_{\partial\mathfrak{G}} \kappa_g\,ds = 2\pi - \sum \delta_i,$$

where the line integral goes over all smooth pieces of the curve, while the summation extends over all corners.

The proof is quite similar to that of Corollary 1.

Theorem 17.3.1 can, in particular, be applied to closed surfaces. Let us consider first closed surfaces obtainable from the full sphere by continuous (and sufficiently differentiable) deformations. Such surfaces are decomposable into two simply connected regions by a closed curve that is piecewise of class C^2. We apply the formula of Gauss and Bonnet to each of these regions and we note that κ_g changes sign if the orientation of the boundary curve is reversed. If the entire closed surface is oriented in a definite way then the decomposing curve, as the common boundary of the two regions, must be traversed once in each of its two opposite orientations. Hence, when we add the Gauss-Bonnet formulas for the two regions, the integrals over the boundary cancel and we have the following result: The total curvature (*curvatura integra*) of a closed surface of the topological type of the sphere is equal to 4π. This total curvature is an example of a topological invariant: It has the same value for every surface that is topologically equivalent to the full sphere.

This result can be generalized. Consider two closed surfaces, each of which is decomposed into finitely many simply connected regions by means of finitely many arcs that are piecewise of class C^2. Assume there exists a topological mapping from one surface onto the other that maps the decomposing arcs of the one onto those of the other. Assume, further, that the surfaces are orientable so that each arc, considered as part of the boundary of two contiguous regions, is traversed in opposite directions as the boundaries of the two regions are traversed in their proper directions. If the Gauss-Bonnet formulas for the separate regions are added up, the line integrals (over their boundaries) will all cancel out. Each corner (or vertex) contributes the sum of the angles at that point, hence a multiple of 2π, and each of the regions contributes 2π to its formula. Thus, the total result of the addition depends only on the combinatorial aspects of the decomposition and is therefore the same for the two closed surfaces. This establishes the following result:

Theorem 17.3.2. *The total curvature* (curvatura integra)

$$\oint\!\!\oint K\,dF$$

of an orientable closed surface is a topological invariant and a multiple of 2π. (See also Fig. 44.)

The second statement of the theorem can be made more specific: We have

$$\oint\!\!\oint K\,dF = 4\pi(1 - p),$$

where the integer p is the *genus* of the surface. For this fact, and for connections to the Euler characteristic, the reader is referred to textbooks of topology and to W. Blaschke, "Einführung in die Differentialgeometrie," p. 53 ff., Berlin-Göttingen-Heidelberg, 1950 (Springer "Grundlehren" series, Vol. 58).

As a simple application of the theorem of Gauss and Bonnet, we can derive a very old theorem (Jacobi, 1842) on curves in the large:

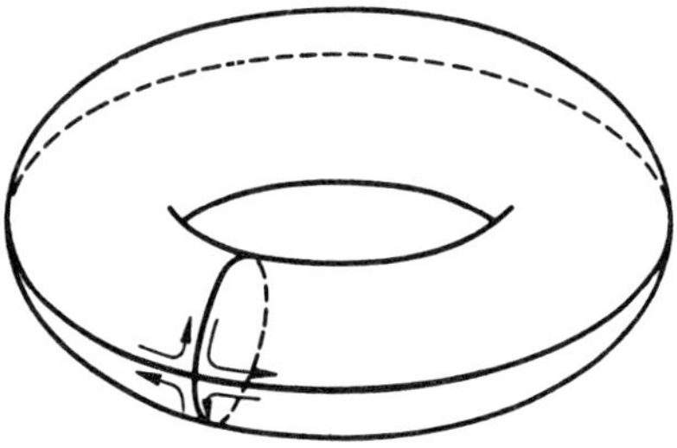

FIG. 44. $\oint\!\!\oint K\,dF = 0$ for the torus.

Theorem 17.3.3. *If the spherical principal normal image of a closed space curve is free of double points, then it divides the surface of the unit sphere into two regions of equal area.*

Proof. Let t be the arc length of the spherical principal normal image $\mathfrak{v}_2$ of the curve $\mathfrak{x}(s)$, and let $\bar{\kappa}_g$ be the geodesic curvature of $\mathfrak{v}_2$ on the unit sphere. Since $K = 1$, we have for the area F of each the spherical regions bounded by $\mathfrak{v}_2(t)$ that $F = \int dF = 2\pi - \oint \bar{\kappa}_g\,dt$, and the conclusion of Theorem 17.3.3 then follows as soon as we have proved the following:

Lemma. *The geodesic curvature $\bar{\kappa}_g$ of $\mathfrak{v}_2$ is given by*

$$\bar{\kappa}_g = \frac{d}{ds}\arctan\frac{\tau}{\kappa}\cdot\frac{ds}{dt},$$

whence

$$\oint \bar{\kappa}_g\,dt = 0.$$

Indeed, by the second of Frenet's formulas, we have

$$\dot{\mathfrak{v}}_2 = \frac{ds}{dt}(-\kappa\mathfrak{v}_1 + \tau\mathfrak{v}_3),$$

and since this must be a unit vector (t being arc length on $\mathfrak{v}_2$), we have

$$\frac{dt}{ds} = \sqrt{\kappa^2 + \tau^2}.$$

Now

$$\begin{aligned} \bar{\kappa}_g &= \mathfrak{v}_2\, \dot{\mathfrak{v}}_2\, \ddot{\mathfrak{v}}_2 \\ &= + \mathfrak{v}_1\, \mathfrak{v}_2\, \mathfrak{v}_3 \left(\frac{ds}{dt}\right)^3 (-\kappa'\tau + \kappa\tau') = \frac{-\kappa'\tau + \kappa\tau'}{\kappa^2+\tau^2}\frac{ds}{dt} \\ &= \frac{d}{ds}\arctan\frac{\tau}{\kappa}\cdot\frac{ds}{dt}, \end{aligned}$$

where we have substituted for derivatives of the $\mathfrak{v}_i$ from the Frenet formulas, and the Lemma and Theorem are proved.

Exercise 17.3.4. Assuming that two geodesics on a surface meet at more than one point, what can be said about the sign of K? (*Hint.* Apply Corollary 2 to the geodesic "lens" formed by the two geodesics.)

17.4 A Condition for the Congruence of Isometric Surfaces

Let $\mathfrak{x}$ and $\tilde{\mathfrak{x}}$ be two isometric surfaces, and assume these to be parametrized in such a way that points that correspond to each other under the isometry have the same parameters. Then $\tilde{g}_{ik} = g_{ik}$ is necessary and sufficient for the mapping to be an isometry.

Two surfaces are said to be *congruent* if each can be obtained from the other by a rigid motion (including reflections, if necessary). If the surfaces are suitably oriented, then, according to the fundamental theorem of §11.7, a necessary and sufficient condition for congruence is that $\tilde{g}_{ik} = g_{ik}$ and $\tilde{L}_{ik} = L_{ik}$.

We now wish to derive a more convenient condition for the congruence of two isometric surfaces. We begin with the necessary and sufficient condition given by the fundamental theorem,

$$A_{ik} \underset{\text{def}}{=} L_{ik} - \tilde{L}_{ik} = 0. \tag{17.22}$$

Now the following result holds:

Theorem 17.4.1. *Let $\mathfrak{x}$ and $\tilde{\mathfrak{x}}$ be two isometric regions on surfaces of class C^3 on which $K \geqq 0$ (i.e., there are no hyperbolic points) and on which the parabolic points are nowhere dense. Then, for the congruence of $\mathfrak{x}$ and $\tilde{\mathfrak{x}}$, the following condition is necessary and sufficient:*

$$A \underset{\text{def}}{=} \det(A_{ik}) = 0. \tag{17.23}$$

(This is a condition of scalar rather than of tensorial type and is therefore easier to apply than $\tilde{L}_{ik} = L_{ik}$.)

Proof. The necessity of the condition follows from the fundamental theorem. To show its sufficiency, we note first that $A = 0$ means $A_{11} A_{22} = (A_{12})^2$, so that the adjoint tensor $*A^{ik} = \varepsilon^{ij} \varepsilon^{kl} A_{jl}$ satisfies $\det(*A^{ik}) = 0$. Therefore, as we know from linear algebra (see also Exercise 17.4.3 below), there exists a vector field a^i such that $a^i a^k = *A^{ik}$, since $*A^{ik}$ is symmetric. If a^i is the null vector then $*A^{ik} = 0$, hence $A_{ik} = 0$, and $\tilde{L}_{ik} = L_{ik}$. Otherwise, we have by (17.12) that

$$\frac{A}{g} = \tfrac{1}{2} *A^{ij} A_{ij} = \tfrac{1}{2}(*L^{ij} - *\tilde{L}^{ij})(L_{ij} - \tilde{L}_{ij}) = *A^{ij} L_{ij}, \tag{17.24}$$

where we have also used (17.13) and the fact that $K = \tilde{K}$, which follows from isometry, and the fact that $*\tilde{L}^{ij} L_{ij} = *L^{ij} \tilde{L}_{ij}$. If we now substitute the decomposition $*A^{ij} = a^i a^j$, it follows that

$$L_{ij} a^i a^j = 0,$$

which, at elliptic points, leads to $a^i = 0$, since $L_{ij} a^i a^j = 0$ is definite at such points. Finally, at parabolic points we consider

$$*A_k{}^k = g_{ik} a^i a^k;$$

this function is continuous and vanishes at all elliptic points, hence also at all parabolic points, since these are, by assumption, accumulation points of elliptic points. Therefore $a^i = 0$ identically, and the theorem is proved.

We note in passing that the validity of the theorem essentially extends also to the hyperbolic case ($K < 0$); a proof for this is to be found in E. Rembs and K. P. Grotemeyer, Appendix to N. W. Efimov, "Flächenverbiegung im Großen," p. 152, Berlin, 1957.

Our next step, toward a use of the method of integral formulas (in §17.5 below), is to prove that A does not change its sign if the surfaces are isometric (but not necessarily congruent):

Theorem 17.4.2. *Let* $\mathfrak{x}$, $\tilde{\mathfrak{x}}$ *be given as in Theorem* 17.4.1, *and assume also that the normals are oriented in such a way that the mean curvatures* H, $\tilde{H}$ *are not negative* (*so that* L_{ik} *and* $\tilde{L}_{ik}$ *are positive semidefinite*). *Then* $A \leqq 0$.

Proof. Consider the quadratic form in x,

$$Q(x) \underset{\text{def}}{=} \frac{\det(L_{ik} - x\tilde{L}_{ik})}{g} = \tfrac{1}{2} ({}^*L^{ik} - x\,{}^*\tilde{L}^{ik})(L_{ik} - x\tilde{L}_{ik})$$
$$= K - Jx + Kx^2,$$

where

$$J \underset{\text{def}}{=} {}^*L^{ik}\,\tilde{L}_{ik} = {}^*\tilde{L}^{ik}\,L_{ik}, \tag{17.25}$$

whence, by (17.24),

$$A/g = 2K - J. \tag{17.26}$$

If $K > 0$, then the forms L_{ik}, $\tilde{L}_{ik}$ are both positive-definite, hence $Q(x) = 0$ has two real roots (therefore $J^2 - 4K^2 \geqq 0$ for the discriminant), and these roots are positive (therefore $J > 0$, by Vieta).

Now, from $K > 0$, $J > 0$, and $0 \leqq J^2 - 4K^2 = (J - 2K)(J + 2K)$, it follows that

$$2K - J \leqq 0,$$

from which the assertion follows, for the case $K > 0$, by (17.26). For $K = 0$ the assertion now follows from the continuity of A and from the fact that every parabolic point is an accumulation point of elliptic points.

Exercise 17.4.3. *Prove*: If $a_{ik} = a_{ki}$ and $\det(a_{ik}) = 0$ $(i, k = 1, 2)$, then there exist a_i such that $a_{ik} = a_i\, a_k$.

Exercise 17.4.4. Extend Theorem 17.4.2 to parabolic regions. [*Hint*. Apply Exercise 17.4.3 to L_{ik} and $\tilde{L}_{ik}$ and substitute into (17.25) and (17.26).]

17.5 Integral Formulas

We shall deduce the congruence of certain pairs of isometric surfaces by showing, for these pairs, that the nonpositive function A of Theorem 17.4.2 vanishes identically. The method of integral formulas will consist, in this case, in finding a function φ of fixed sign such that, under suitable assumptions, the integral of $\varphi \cdot A$ vanishes, which then implies $A = 0$.

We begin by writing down the formula of Gauss for the derivatives,

$$\mathfrak{x}_{i;k} = L_{ik}\,\mathfrak{N},$$

as well as the formula that is obtained, on account of (17.22), from the Mainardi-Codazzi equations (17.14) of the two given isometric surfaces $\mathfrak{x}$, $\tilde{\mathfrak{x}}$:

$$*A^{ik}{}_{;k} = 0. \tag{17.27}$$

We also have, from (17.24),

$$\frac{A}{g} = *A^{ik} L_{ik}. \tag{17.28}$$

By inner multiplication of Gauss' equations with $*A^{ik}$, we therefore obtain

$$(*A^{ik} \mathfrak{x}_i)_{;k} = \frac{A}{g} \mathfrak{N}. \tag{17.29}$$

From this formula we can obtain integral formulas (see Rembs and Grotemeyer, *loc. cit.*) after multiplying by a function G (which may also be a vector function, in which case inner multiplication is called for):

$$\iint_{\mathfrak{G}} \frac{A}{g} \mathfrak{N} G \, dF = -\oint_{\partial\mathfrak{G}} *A^{ik} G \mathfrak{x}_i h_k \, ds - \iint_{\mathfrak{G}} *A^{ik} G_{,k} \mathfrak{x}_i \, dF. \tag{17.30}$$

Here, the integral arising from the left-hand side of (17.29) has been transformed by integration by parts. Also, the boundary $\partial\mathfrak{G}$ of our region is assumed to be of class C^2, and h_k is defined as in §17.2. Depending on the choice of $\mathfrak{G}$, various integral formulas are obtained. For instance, choosing for G the position vector $\mathfrak{x}$, we obtain

$$\iint_{\mathfrak{G}} \mathfrak{N}\mathfrak{x} \frac{A}{g} \, dF = -\oint_{\partial\mathfrak{G}} *A^{ik} \mathfrak{x}\mathfrak{x}_i h_k \, ds - 2\iint_{\mathfrak{G}} (H - \tilde{H}) \, dF. \tag{17.31}$$

The mean curvatures H, $\tilde{H}$ enter on account of $\mathfrak{x}_k \mathfrak{x}_i = + g_{ik}$ and $H = \frac{1}{2} g^{ik} L_{ik} = \frac{1}{2} *L^{ik} g_{ik}$ and $g_{ik} = \tilde{g}_{ik}$. Formula (17.31) is due to Herglotz. We can get rid of the double integral on the right by interchanging the roles of $\mathfrak{x}$ and $\tilde{\mathfrak{x}}$. If we add the resulting formula to (17.31), we obtain the important symmetric from of Herglotz' integral formula,

$$\iint_{\mathfrak{G}} \frac{A}{g} (\mathfrak{N}\mathfrak{x} + \tilde{\mathfrak{N}} \tilde{\mathfrak{x}}) \, dF = \text{boundary integral}. \tag{17.32}$$

Here, $\mathfrak{N}\mathfrak{x}$ and $\tilde{\mathfrak{N}}\tilde{\mathfrak{x}}$ are the Minkowski support functions of the two surfaces. Except for their sign, they are equal to the distance of the tangent plane from the origin, being the projections of the position vectors onto the unit normals to the surfaces.

The same method can be used to derive other formulas. If we (inner) multiply the formulas of Gauss with g^{ik}, say, then we obtain

$$(g^{ik}\,\mathfrak{x}_i)_{;k} = 2H\mathfrak{N} \tag{17.33}$$

and from this, just as above with $G = \mathfrak{x}$, we obtain the integral formula of H. Minkowski for closed surfaces (so that the integral over the boundary falls away):

$$\iint_{\mathfrak{G}} H\mathfrak{N}\mathfrak{x}\,dF = -F, \tag{17.34}$$

where F denotes the area of the entire surface. If we (inner)multiply the formulas of Gauss with $^*L^{ik}$ and make use of the Mainardi-Codazzi equations (17.14) as well as of formula (17.13), we obtain similarly

$$(^*L^{ik}\,\mathfrak{x}_i)_{;k} = 2K\mathfrak{N} \tag{17.35}$$

and from this, for the case of closed surfaces, a further formula of Minkowski:

$$\iint_{\mathfrak{G}} K\mathfrak{N}\mathfrak{x}\,dF = -\iint_{\mathfrak{G}} H\,dF. \tag{17.36}$$

17.6 The Congruence of Isometric Ovaloids

After these preparations and with the use of the integral formula of Herglotz, it is now easy to present the proof of Herglotz (1943) for the following theorem (of St. Cohn-Vossen, 1927):

Theorem 17.6.1. *If two ovaloids of class C^3 are isometric, and if $K \geqq 0$ throughout, with the parabolic points being nowhere dense, then the ovaloids are congruent; that is, they can be transformed into each other by a rigid motion (including a reflection, if necessary).*

Proof. If necessary, we first carry out a translation to insure that the interiors of the two ovaloids have points in common and we choose one of these points as the coordinate origin. In accordance with making both L_{ik} and $\tilde{L}_{ik}$ positive semidefinite forms, we orient the surfaces in such a way that their normals point into their interiors; then the support functions $\mathfrak{x}\mathfrak{N}$ and $\tilde{\mathfrak{x}}\tilde{\mathfrak{N}}$ are always negative. Since $A \leqq 0$ by Theorem 17.4.2, the integrand in the Herglotz formula (17.32) is always nonnegative. The boundary integral vanishes, since the surfaces are closed; hence the integrand on the left must vanish identically. This implies that $A \equiv 0$, and now Theorem 17.4.1 completes the proof.

We note in passing that the differentiability conditions in Theorem 17.6.1 can be weakened, and that the conclusion still holds even if subregions are admitted that consist entirely of parabolic points. Details and further references can be found in the report by Rembs and Grotemeyer cited in §17.4.

Theorem 17.6.1 contains in particular the result, first proved by H. Liebmann in 1900, that ovaloids cannot be bent: *A continuous* (and sufficiently differentiable) *deformation of an ovaloid that does not change any of its intrinsic measurements must be a rigid motion of the whole surface.* A somewhat different formulation, which we shall not prove here, is the statement that *ovaloids are rigid,* which was also proved by Liebmann. A surface is called rigid if it does not allow infinitesimal bendings; a continuous deformation is called an infinitesimal bending if any changes in length under the deformation are of higher than the first order. Evidently, if a surface is rigid, then it cannot be bent.

From the intrinsic-geometric view, an ovaloid is a compact two-dimensional Riemannian space with $K > 0$ (or, somewhat more generally, with $K \geqq 0$ and with the points where $K = 0$ being nowhere dense).

In 1916, H. Weyl posed the problem as to whether every such Riemannian space can be realized by an ovaloid. Following Weyl's own attacks on this problem, a great number of mathematicians have succeeded in lowering more and more the analytic conditions under which the question can be answered in the positive. Thus it is known that every compact two-dimensional Riemannian space of positive curvature can be represented by an ovaloid, uniquely to within rigid motions (by Theorem 17.6.1).

APPENDIX I

From the History of Differential Geometry

Differential geometry builds on the following disciplines as its prerequisites: The analytic geometry of Descartes (1596–1650) and the calculus (Leibniz, 1646–1716; Newton, 1643–1727). The first, isolated results on curves and surfaces date from the 18th century and here, as in the other branches of analysis, much was contributed especially by L. Euler (1707–1783). In France, during its Revolution and First Empire periods, G. Monge (1746–1818) founded an extensive school of geometry which enriched differential geometry with many essential contributions. Monge's pupils included C. Dupin (1784–1873; "Développements de géométrie," 1813). Geometry in France was under the influence of the tradition of Monge's school even into the 20th century.

C. F. Gauss (1777–1855) transformed the theory of surfaces into its modern systematic mold. He recognized the fundamental significance of intrinsic geometry. His main work in differential geometry is his treatise of 1827, *Disquisitiones generales circa superficies curvas*. A foundation of intrinsic geometry independent of imbedding was given by B. Riemann (1826–1866) in his inaugural address, delivered at Göttingen in 1854 in the presence of the elder Gauss (*Über die Hypothesen, welche der Geometrie zugrunde liegen*). There he also dropped the restriction to two dimensions and laid the foundation for "Riemannian geometry" which has since been extensively developed, partly because of its applications in physics.

Riemann's contributions were not published until 1868, after his death. The further development was at first influenced, even more than by Riemann's ideas, by Felix Klein (1849–1925) who in his "Erlangen program" of 1872 defined a geometry as being the theory of invariants of a group of transformations. He had in mind particularly the group of projective transformations and its subgroups. Thus he sees Euclidean geometry as the theory of invariants of the group of rigid motions.

Around the turn of the 20th century, the tensor calculus was developed as a powerful tool for differential geometry [G. Ricci and T. Levi-Cività, Méthodes de calcul différentiel absolu et leurs applications, *Math. Ann.* **54** (1901)]. Together with the general theory of relativity of A. Einstein

[1879–1955; Zur Allgemeinen Relativitätstheorie, *Sitzber. preuss. Akad. Wiss.* (1915)], this signaled an intensive development of Riemannian geometry and of other geometric structures in differentiable manifolds.

Apart from this, differential geometry followed, in Central Europe of the first half of this century, two lines of development in particular, mainly under the influence of W. Blaschke (1885–1962) and his school. On the one hand, the differential geometry of the groups of the "Erlangen program" was investigated systematically.[1] On the other hand, a great deal of attention was being given to questions of differential geometry in the large, questions into which enter problems concerning the bending of surfaces and the connections with the theory of convex bodies.[2] A new direction was opened up by the study of fiber spaces, which in the last decades has led to new possibilities in topology as well as in differential geometry in the large. For the later field, another tool that has proven its usefulness is a refinement of the tensor calculus (differential forms), which were first used systematically by E. Cartan (1869–1949).

A subject which, from its very beginning, was closely linked to differential geometry is the calculus of variations. In 1918 P. Finsler (b. 1894), following

---

[1] For affine differential geometry, see W. Blaschke, "Vorlesungen über Differentialgeometrie II," Berlin, 1923. For projective differential geometry, see G. Bol, "Projektive Differentialgeometrie I–III," Göttingen, from 1950 on. Also, W. Blaschke, "Vorlesungen über Differentialgeometrie III," and E. Salkowski, "Affine Differentialgeometrie," Berlin and Leipzig, 1934.

Notwithstanding the fruitfulness of Klein's Erlangen program, it should be pointed out that there is a danger in considering this program as a regulative principle of geometry altogether (which is still often done, consciously or unconsciously), which would lead to an *a priori* rejection of many parts of differential geometry as being less "geometrical." Thus Riemann's geometry, discovered before Klein, could only artificially be subsumed under the Erlangen program, while metric geometry — which proceeds purely from properties of a distance and does not immediately make any use of differentiability properties — would lie entirely outside the Erlangen concept of a geometry.

If Klein's program is, on the one hand, too narrow for our present-day conception of geometry, it is true, on the other hand, that views have changed even within the original domain of validity of Klein's ideas. The geometry of a (nonmetrized) vector space is, in Klein's view, the geometry of the group of homogeneous substitutions; but, for us today, this group of automorphisms is only a secondary feature of a vector space. This criticism does, of course, not touch the importance of the role that was played by the Erlangen program in the history of geometry.

[2] See, for instance, N. W. Efimov, "Flächenverbiegung im Großen," with an appendix by E. Rembs and K. P. Grotemeyer, Berlin, 1957; also the classical works of W. Blaschke: "Kreis und Kugel," 1916; 2nd ed., Berlin, 1956; reprint, New York, 1949; and "Vorlesungen über Differentialgeometrie I," 4th ed., Berlin, 1945.

suggestions by C. Carathéodory (1873–1949), wrote his Göttingen dissertation ("Über Kurven und Flächen in allgemeinen Räumen"; new printing, Basel, 1951) in which this old connection was used to construct a new metric differential geometry that has since been developed considerably. From this new vantage point one can gain a new understanding of Euclidean, non-Euclidean and Riemannian geometries, an understanding pioneered by the contributions (1868) of H. von Helmholtz (1821–1894).

Finally we mention recent investigations of problems that are, at least in part, of differential-geometric interest, by coordinate-free methods not based on any special analytic assumptions, such as are found in A. D. Aleksandrov, "Die innere Geometrie der konvexen Flächen," Berlin, 1955, and H. Busemann, "The Geometry of Geodesics," New York, 1955. Busemann has recently proved that under certain assumptions that can be formulated in an intrinsic-geometric fashion, the metric spaces basic to his investigations are Finsler spaces; thus his line of inquiry also appears to lead back into differential geometry. We cannot enter in this book into the close connections that link differential geometry with the theory of Lie groups.

APPENDIX II

Some Prerequisite Theorems of Analysis

1. Existence and Uniqueness Theorem for Ordinary First-Order Differential Equations and Systems of Equations

For the proof of the following theorem, the reader is referred to textbooks on ordinary differential equations.

Existence and Uniqueness Theorem. *Let there be given a system of differential equations*

$$\frac{dx_i}{dt} = f_i(t; x_1, \ldots, x_n) = f_i(t; x_k) \qquad (n \geqq 1);$$

assume that the functions f_i are defined and continuous in a region R of the $(n+1)$-dimensional space of $(t; x_i)$ and that these functions satisfy in R a Lipschitz condition (L: a constant):

$$\sum_i |f_i(t; x_k') - f_i(t; x_k)| \leqq L \sum_i |x_i' - x_i|.$$

(If the f_i have continuous and bounded partial derivatives in R with respect to the x_k, then such a Lipschitz condition is certainly satisfied in R, by the Law of the Mean.)

Then corresponding to initial conditions $x_i(t_0) = \overset{0}{x}_i$ in R, there exists in some interval $|t - t_0| < a$ one and only one solution $x_i = x_i(t)$ of the system, satisfying $x_i(t_0) = \overset{0}{x}_i$. If the f_i are of class C^r, then the solution is of class C^{r+1}; if the f_i are analytic, then so is the solution. If the f_i depend continuously (or of class C^r, with $r \geqq 1$) on parameters ε_ρ, then the solution depends continuously (or of class C^r, resp.) on these parameters. The ε_ρ may, for instance, be the assigned initial values.

Systems of the second or higher order can be reduced to first-order systems. For example, if

$$\frac{d^2x_i}{dt^2} = f_i(t; x_k; \dot{x}_k), \qquad i, k = 1, \ldots, n$$

is the system (of second order) to be solved, we set $\dot{x}_i \underset{\text{def}}{=} x_{i+n}$ and we thus obtain a 1st-order system for the $2n$ variables $x_1, \ldots, x_{2n}$. As initial conditions determining a unique solution, we can assign at t_0 the values of $x_1, \ldots, x_{2n}$, that is, of

$$x_1, \ldots, x_n; \quad \dot{x}_1, \ldots, \dot{x}_n.$$

2. Integrability Theory for Systems of First-Order Partial Differential Equations

Theorem. *Let there be given the following system of first-order partial differential equations:*

$$\frac{\partial y^i}{\partial x^\alpha} = f^i_\alpha(x^\beta; y^k). \tag{1}$$

(Italic indices range from 1 to n; Greek indices from 1 to m.) Assume the functions f^i to be of class C^2 in a region R of their $(m+n)$-dimensional argument space.

The system (1) is said to be completely integrable if, for every point $(\underset{0}{x}^\beta; \underset{0}{y}^k)$ of R, there exists at least one solution $y^i(x^\alpha)$ of (1), of class C^2, that satisfies the initial condition $y^i(\underset{0}{x}^\alpha) = \underset{0}{y}^i$.

Then the following two conclusions hold:

(a) *Uniqueness: There is at most one solution satisfying a given initial condition.*

(b) *Existence: The system is completely integrable if and only if the following integrability condition holds in R:*

$$J^i_{\alpha\beta} = 0, \tag{2}$$

where

$$J^i_{\alpha\beta} \underset{\text{def}}{=} f^i_{\alpha,\beta} - f^i_{\beta,\alpha} + f^i_{\alpha,k} f^k_\beta - f^i_{\beta,k} f^k_\alpha \tag{3}$$

($,\beta$ and $,k$ denote partial differentiations with respect to the corresponding variables.)

Proof. We prove first the necessity of (2). If $y^i(x^\alpha)$ is a solution of class C^2 satisfying $y^i(\underset{0}{x}^\alpha) = \underset{0}{y}^i$, then, from the equality of the second mixed partials, it follows that $y^i_{,\alpha,\beta} - y^i_{,\beta,\alpha} = 0$. Using the chain rule and the given differential equations (1), we see that this amounts to $J^i_{\alpha\beta} = 0$, at the point $(\underset{0}{x}^\alpha, \underset{0}{y}^i)$. This point may be any point of R and the necessity of the condition is established.

We still have to prove (a), and the sufficiency of (2) for (b). Consider the following system of ordinary differential equations:

$$\frac{dy^i}{dt} = f^i_\alpha(x^\beta(t); y^k) \frac{dx^\alpha}{dt}. \tag{4}$$

Since $dy^i/dt = (\partial y^i/\partial x^\alpha)(dx^\alpha/dt)$, every solution $y^i(x^\beta)$ of (1), evaluated along the curve $x^\beta(t)$ [that is, with $y^i(t) \underset{\text{def}}{=} y^i(x^\beta(t))$], must be a solution of (4). By the theorem of Appendix II.1, there is one and only one solution $y^i(t)$ such that $y^i(0) = \underset{0}{y}^i$ for $x^\beta(0) = \underset{0}{x}^\beta$ and we have therefore obtained the uniqueness conclusion (a): Any two solutions belonging to the same initial conditions must have the same values at any point (joined to $\underset{0}{x}^\beta$ by a curve) and must therefore coincide altogether.

In order to show the existence of a solution we must use (2). It is sufficient to prove the following: If this integrability condition holds, then the integration of (4) along different paths $x^\beta(t)$, $\bar{x}^\beta(t)$ satisfying $x^\beta(0) = \bar{x}^\beta(0) = \underset{0}{x}^\beta$ and

$$x^\beta(1) = \bar{x}^\beta(1) = \underset{1}{x}^\beta$$

leads to the same value $y^i(\underset{1}{x}^\beta)$, provided that $y^i(0) = \underset{0}{y}^i$ is the initial condition in both cases. This will give us a function $y^i(x^\beta)$ uniquely defined in a neighborhood of $\underset{0}{x}^i$ and satisfying the system (1).

We consider all paths $x^\beta(t;\varepsilon) = (1-\varepsilon)x^\beta(t) + \varepsilon\bar{x}^\beta(t)$, $0 \leqq \varepsilon \leqq 1$; these are all of class C^2, if x^β and $\bar{x}^\beta$ are of class C^2, and they satisfy $x^\beta(0;\varepsilon) = \underset{0}{x}^\beta$, $x^\beta(1;\varepsilon) = \underset{1}{x}^\beta$. We denote by $y^i(t;\varepsilon)$ the corresponding solutions of (4) that satisfy

$$y^i(0;\varepsilon) = \underset{0}{y}^i.$$

Since the right-hand side of (4) is a differentiable function of the parameter ε, so is the function $y^i(t;\varepsilon)$, by Appendix II.1. We can therefore construct

$$z^i(t;\varepsilon) \underset{\text{def}}{=} \frac{dy^i}{d\varepsilon} - f^i_\alpha(x(t;\varepsilon); y(t;\varepsilon)) \frac{dx^\alpha}{d\varepsilon}. \tag{5}$$

We differentiate this with respect to t and substitute for dy/dt from (4), obtaining

$$\frac{dz^i}{dt} = f^i_{\alpha,k}\frac{dy^k}{d\varepsilon}\frac{dx^\alpha}{dt} + f^i_{\alpha,\beta}\frac{dx^\beta}{d\varepsilon}\frac{dx^\alpha}{dt} - f^i_{\alpha,k}\frac{dy^k}{dt}\frac{dx^\alpha}{d\varepsilon} - f^i_{\alpha,\beta}\frac{dx^\beta}{dt}\frac{dx^\alpha}{d\varepsilon}$$

or

$$\frac{dz^i}{dt} = f^i_{\alpha,k}\, z^k \frac{dx^\alpha}{dt} + J^i_{\alpha\beta}\frac{dx^\alpha}{dt}\frac{dx^\beta}{d\varepsilon}\,. \tag{6}$$

By hypothesis (2), the last term on the right vanishes. Since $z^i(0;\varepsilon) = 0$ by the initial conditions, it now follows from the uniqueness theorem that $z^i(t;\varepsilon) = 0$ is the only solution of (6). Since $dx^\beta/d\varepsilon = 0$ at $t = 1$, it follows from (5) that $dy^i/d\varepsilon = 0$ at $t = 0$, so that $y^i(\underset{1}{x})$ is indeed independent of the path used to join $\underset{0}{x}^\beta$ to $\underset{1}{x}^\beta$, q.e.d.

This method for reducing problems about systems of partial differential equations to well-known facts about ordinary systems is of quite general applicability. It can be used, for instance, to prove statements concerning the dependence of the solutions on the initial conditions.

3. From the Calculus of Variations

Fundamental Lemma of the Calculus of Variations. *Let R be a region of n-dimensional space, $n \geqq 1$. (For $n = 1$, R is an interval and its boundary ∂R consists of its two end points.) Let f be a function continuous in R and such that for every function g of class C^∞ in R and vanishing on ∂R, the integral of $f \cdot g$ over R vanishes. Then f vanishes identically in R.*

Proof. Otherwise, there would be a point $\underset{0}{x}$ in R and a neighborhood $\Sigma(x^i - \underset{0}{x}^i)^2 < \rho^2$ where $f \neq 0$, say $f > 0$. Then the function

$$g(x^i) = \begin{cases} \exp\left(-1/(\rho^2 - r^2)\right) & \text{for} \quad r^2 \underset{\text{def}}{=} \sum (x^i - \underset{0}{x}^i)^2 < \rho^2 \\ 0 & \text{otherwise} \end{cases}$$

would be of class C^∞ and the integral of $f \cdot g$ over R would be positive, contradicting one of the hypotheses.

We shall now derive the differential equations of Euler and Lagrange for parameter-dependent variational problems of the following type: Let

$F(x^i; \dot{x}^i)$ be of class C^2 in all of its variables, and let $x^i = x^i(t)$, $t_0 \leqq t \leqq t_1$, be a curve of class C^2 that yields an extremum for the integral $\int_{t_0}^{t_1} F\,dt$ by comparison with all curves $x^i(t) + \varepsilon y^i(t)$ satisfying $y^i(t_0) = y^i(t_1) = 0$, the function $y^i(t)$ otherwise being an arbitrary function, say, of class C^∞. Since the limit operations commute on account of uniformity properties, we have

$$0 = \frac{d}{d\varepsilon} \int_{t_0}^{t_1} F(x + \varepsilon y; \dot{x} + \varepsilon \dot{y})\,dt \Bigg|_{\varepsilon=0} = \int_{t_0}^{t_1} \left(\frac{\partial F}{\partial x^i} y^i + \frac{\partial F}{\partial \dot{x}^i} \dot{y}^i \right) dt.$$

We integrate the last integral by parts. Some of the resulting terms vanish on account of $y^i(t_0) = y^i(t_1) = 0$, and we obtain

$$\int_{t_0}^{t_1} \left(\frac{\partial F}{\partial x^i} - \frac{d}{dt} \frac{\partial F}{\partial \dot{x}^i} \right) y^i\,dt = 0.$$

In particular, if we set $y^i = 0$ for $i \neq k$ and apply the fundamental lemma, we obtain the following differential equations as necessary conditions for an extremum:

$$\frac{\partial F}{\partial x^k} - \frac{d}{dt} \frac{\partial F}{\partial \dot{x}^k} = 0.$$

APPENDIX III

Summary of Formulas

Theory of Curves

Curve: $\mathfrak{x} = \mathfrak{x}(t)$ in Euclidean 3-space. Arc length s: $ds = |\dot{\mathfrak{x}}|\, dt = \sqrt{\dot{\mathfrak{x}}^2}\, dt$, hence $\mathfrak{x}'^2 = 1$. (Prime denotes derivative with respect to arc length s) (§1.2).

Moving trihedron: consists of unit tangent vector $\mathfrak{v}_1 = \mathfrak{x}'$, (unit) principal normal vector $\mathfrak{v}_2 = \mathfrak{x}''/|\mathfrak{x}''|$, and (unit) binormal vector $\mathfrak{v}_3$ (§1.4).

Frenet-Serret formulas (§1.5):

$$\begin{aligned} \mathfrak{v}_1' &= \quad\quad\quad \kappa\mathfrak{v}_2 \\ \mathfrak{v}_2' &= -\kappa\mathfrak{v}_1 \quad\quad\quad + \tau\mathfrak{v}_3 \\ \mathfrak{v}_3' &= \quad\quad -\tau\mathfrak{v}_2. \end{aligned}$$

Curvature: $\kappa = |\mathfrak{x}''|$

Torsion: $\tau = \mathfrak{v}_2'\,\mathfrak{v}_3 = \mathfrak{x}'\,\mathfrak{x}''\,\mathfrak{x}'''/(\mathfrak{x}'')^2$ (Excercise 1.7.1).

Theory of Surfaces

Surface in parametric representation: $\mathfrak{x} = \mathfrak{x}(u^1, u^2) = \mathfrak{x}(u^i)$; $\mathfrak{x}_i \underset{\text{def}}{=} \partial\mathfrak{x}/\partial u^i$.
(Unit) normal vector: $\mathfrak{N} = (\mathfrak{x}_1 \times \mathfrak{x}_2)/|\mathfrak{x}_1 \times \mathfrak{x}_2|$.
Coefficients of first fundamental form (fundamental metric tensor):

$$g_{ik} = \mathfrak{x}_i\,\mathfrak{x}_k \qquad (§3.4).$$

Coefficients of second fundamental form (second fundamental tensor):

$$L_{ik} = \mathfrak{N}\mathfrak{x}_{ik} = -\mathfrak{N}_i\,\mathfrak{x}_k = -\mathfrak{N}_k\,\mathfrak{x}_i \qquad (§3.5).$$

Coefficients of the ε-tensor ε_{ik}:

$$\varepsilon_{11} = \varepsilon_{22} = 0, \qquad \varepsilon_{12} = -\varepsilon_{21} = \sqrt{g}$$

$$g = \det(g_{ik}) > 0, \qquad g^{ik}\,g_{kj} = \delta^i_j, \qquad \delta^i_j = \begin{cases} 0 & \text{for} \quad i \neq j \\ 1 & \text{for} \quad i = j \end{cases}$$

$$L = \det(L_{ik}) \qquad (§4.2).$$

Christoffel symbols:

$$\Gamma_{ik|l} = \mathfrak{x}_{ik}\,\mathfrak{x}_l = \tfrac{1}{2}\,(g_{il,k} - g_{ik,l} + g_{kl,i}), \quad g_{il,k} = \frac{\partial g_{il}}{\partial u^k}, \qquad \Gamma^j_{ik} = g^{jl}\,\Gamma_{ik|l} \qquad (\S 4.2).$$

Equations for derivatives, by

Gauss: $\mathfrak{x}_{ik} = \Gamma^r_{ik}\,\mathfrak{x}_r + L_{ik}\,\mathfrak{N}$

$\mathfrak{x}_{i;k} = L_{ik}\,\mathfrak{N}$ (; for covariant derivative)

Weingarten: $\mathfrak{N}_i = -\,L^r_i\,\mathfrak{x}_r, \qquad L^r_i = g^{rj}\,L_{ji}.$

Curvature vector of a curve $\mathfrak{x}(s) = \mathfrak{x}(u^i(s))$ on the surface:

$$\mathfrak{x}'' = \kappa_g\,\mathfrak{s} + \kappa_n\,\mathfrak{N}, \qquad \mathfrak{s} = \mathfrak{N} \times \mathfrak{x}'.$$

Geodesic curvature:

$$\kappa_g = \mathfrak{N}\mathfrak{x}'\,\mathfrak{x}'' = \kappa\mathfrak{N}\mathfrak{v}_3 = \varepsilon_{li}\,u^{l'}(u^{i''} + \Gamma^i_{jk}\,u^{j'}\,u^{k'}). \qquad (\S 4.2)$$

Geodesic curves:

$$\kappa_g = 0 \qquad \text{or} \qquad u^{i''} + \Gamma^i_{jk}\,u^{j'}\,u^{k'} = 0.$$

Normal curvature:

$$\kappa_n = L_{ik}\,\dot{u}^i\,\dot{u}^k / g_{ik}\,\dot{u}^i\,\dot{u}^k$$

(The extrema of κ_n are the principal curvatures κ_1 and κ_2) (§5.1).
Mean curvature:

$$H = \frac{\kappa_1 + \kappa_2}{2} = \frac{1}{2}\,g^{ik}\,L_{ik} \qquad (\S 5.1).$$

Gaussian curvature:

$$K = \kappa_1 \cdot \kappa_2 = L/g \qquad (\S 5.1).$$

Weingarten's equations in line-of-curvature parameters (so-called formulas of Rodrigues):

$$\mathfrak{N}_i = -\,\kappa_i\,\mathfrak{x}_i \qquad \text{(Exercise 5.3.2)}.$$

Parallel vector field (of geodesic parallels) $\xi^i(t)$ along the curve $u^i(t)$:

$$\frac{d\xi^i}{dt} + \Gamma^i_{jk}\,\xi^j\,\frac{du^k}{dt} = 0 \qquad (\S 4.7).$$

Riemannian curvature tensor (§5.4):

$$R^l{}_{ikj} = \Gamma^l{}_{ik,j} - \Gamma^l{}_{ij,k} + \Gamma^r_{ik}\,\Gamma^l_{rj} - \Gamma^r_{ij}\,\Gamma^l_{rk}$$

$$R_{hikj} = g_{hl}\,R^l{}_{ikj}.$$

Theorema egregium:

$$K = \frac{g_{2r}}{g}\,R^r{}_{112} = \frac{1}{g}\,R_{2112} = -\frac{1}{4}\,\varepsilon^{ij}\,\varepsilon^{kl}\,R_{ijkl}.$$

Index

R

S

T